Vilhjalmur Stefansson
1879–1962

Tom Henighan

Tom Henighan is Professor Emeritus at Carleton University, and a faculty member at the Pari Centre for New Learning in Tuscany, Italy. For many years he taught a course in mythology in Carleton's elite Humanities program, and is the author of fifteen published books, several of which have received prize nominations, as well as "notable books" citations in the *Globe and Mail*, and elsewhere. He has been a consultant to many of Canada's major arts institutions, and in 2008 received the Victor Tolgesy Award for lifetime contributions to the arts in Ottawa.

In the same collection

Ven Begamudré, *Isaac Brock: Larger Than Life*
Lynne Bowen, *Robert Dunsmuir: Laird of the Mines*
Kate Braid, *Emily Carr: Rebel Artist*
Kathryn Bridge, *Phyllis Munday: Mountaineer*
William Chalmers, *George Mercer Dawson: Geologist, Scientist, Explorer*
Anne Cimon, *Susanna Moodie: Pioneer Author*
Deborah Cowley, *Lucille Teasdale: Doctor of Courage*
Gary Evans, *John Grierson: Trailblazer of Documentary Film*
Judith Fitzgerald, *Marshall McLuhan: Wise Guy*
lian goodall, *William Lyon Mackenzie King: Dreams and Shadows*
Stephen Eaton Hume, *Frederick Banting: Hero, Healer, Artist*
Naïm Kattan, *A.M. Klein: Poet and Prophet*
Betty Keller, *Pauline Johnson: First Aboriginal Voice of Canada*
Heather Kirk, *Mazo de la Roche: Rich and Famous Writer*
Vladimir Konieczny, *Glenn Gould: A Musical Force*
Michelle Labrèche-Larouche, *Emma Albani: International Star*
Wayne Larsen, *A.Y. Jackson: A Love for the Land*
Wayne Larsen, *James Wilson Morrice: Painter of Light and Shadow*
Francine Legaré, *Samuel de Champlain: Father of New France*
Margaret Macpherson, *Nellie McClung: Voice for the Voiceless*
Nicholas Maes, *Robertson Davies: Magician of Words*
Dave Margoshes, *Tommy Douglas: Building the New Society*
Marguerite Paulin, *René Lévesque: Charismatic Leader*
Marguerite Paulin, *Maurice Duplessis: Powerbroker, Politician*
Raymond Plante, *Jacques Plante: Behind the Mask*
T.F. Rigelhof, *George Grant: Redefining Canada*
Tom Shardlow, *David Thompson: A Trail by Stars*
Arthur Slade, *John Diefenbaker: An Appointment with Destiny*
Roderick Stewart, *Wilfrid Laurier: A Pledge for Canada*
Sharon Stewart, *Louis Riel: Firebrand*
André Vanasse, *Gabrielle Roy: A Passion for Writing*
John Wilson, *John Franklin: Traveller on Undiscovered Seas*
John Wilson, *Norman Bethune: A Life of Passionate Conviction*
Rachel Wyatt, *Agnes Macphail: Champion of the Underdog*

Vilhjalmur Stefansson

Editor: Michael Carroll
Copy Editor: Allison Hirst
Index: Darcy Dunton
Design: Courtney Horner
Printer: Webcom
Cover photo courtesy of Dartmouth College Library

Library and Archives Canada Cataloguing in Publication

Henighan, Tom
     Vilhjalmur Stefansson : Arctic adventurer / by Tom Henighan.

Includes index.
ISBN 978-1-55002-874-4

     1. Stefansson, Vilhjalmur, 1879-1962. 2. Arctic regions--Discovery and exploration--Canadian. 3. Explorers--Canada--Biography. 4. Anthropologists--Canada--Biography. I. Title.

G635.S7H45 2009   917.1904'2092   C2008-906217-5

1   2   3   4   5   13   12   11   10   09

Conseil des Arts
du Canada

Canada Council
for the Arts

ONTARIO ARTS COUNCIL
CONSEIL DES ARTS DE L'ONTARIO

Canadä

We acknowledge the support of the **Canada Council for the Arts** and the **Ontario Arts Council** for our publishing program. We also acknowledge the financial support of the **Government of Canada** through the **Book Publishing Industry Development Program** and **The Association for the Export of Canadian Books**, and the **Government of Ontario** through the **Ontario Book Publishers Tax Credit program**, and the **Ontario Media Development Corporation**.

Printed and bound in Canada.
Printed on recycled paper.

www.dundurn.com

Dundurn Press
3 Church Street, Suite 500
Toronto, Ontario, Canada
M5E 1M2

Gazelle Book Services Limited
White Cross Mills
High Town, Lancaster, England
LA1 4XS

Dundurn Press
2250 Military Road
Tonawanda, NY
U.S.A. 14150

TOM HENIGHAN

Vilhjalmur Stefansson

ARCTIC ADVENTURER

THE QUEST LIBRARY

DUNDURN PRESS
TORONTO

To the Millikens:
Patricia, Bill, Sarah, Erin, Emily, and Andrew,
with thanks for long friendship, generous hospitality,
and many delightful memories of town and country.

Contents

"Exploration is but the physical expression of the intellectual passion."

— Apsley Cherry-Garrard

"I may say that this is the greatest factor — the way in which the expedition is equipped — the way in which every difficulty is foreseen, and precautions taken for meeting or avoiding it. Victory awaits him who has everything in order — luck, people call it. Defeat is certain for him who has neglected to take the necessary precautions in time; this is called bad luck."

— Roald Amundsen

"Exploration is the poetry of action."

— Vilhjalmur Stefansson

Preface

This short biography of the explorer Vilhjalmur Stefansson is designed to provide a readable and concise summary of his life and a fresh estimate of his considerable achievements, with a special emphasis on his Canadian adventures. In terms of research, my book is deeply indebted to several diligent Stefansson biographers and chroniclers, all of whom are credited in the bibliography and cited in the text. I learned from them even when I found it necessary to differ from them.

My book was commissioned as part of a series in which "creative non-fiction" is encouraged, and you will find this approach predominant in the first chapter and in the Epilogue of my book. This kind of writing — using fictional techniques to deal with history or biography — is not new. The wonderful English writer Ford Madox Ford, for example, espoused it — although he called it "impressionism." When Ford wanted to present "the real" Joseph Conrad, or Henry James, or Ernest Hemingway, he did not simply recount biographical facts; he presented his fellow writers, living and breathing, on the page. This approach underlies the form of my book: the first and last sections are meant to reveal by evoking rather than by chronicling. Yet most of the allusions in Chapter One and in my Epilogue, "creative non-fiction" or not, are quite faithful to our information about Stefansson. His dialogue with his dog Dekoraluk, for example, is taken almost verbatim from Stefansson's diary. I do assume in the Epilogue that Stefansson may have told his lover Fanny Hurst about his Inuit family. There's no evidence for that, but it seems perfectly possible to me. Fanny was an enlightened woman and could have handled it; it may also have titillated her. Apart from that, both the imaginative parts of the book and the more mundane sections are as accurate as I can make them. My chief aim is to present Stefansson's life and career as straightforwardly as possible, but also to evoke his remarkable personality and to relate his work to our contemporary ideas of exploration, as well as to traditional and current visions of "the North."

I never met Stefansson, but he is a strong personality and comes across vividly in most of the accounts of him. I'm sure he would be very happy to know that Canadians, at last, are thinking seriously about their most northern regions. He would have been surprised, however, that our contemporary thoughts have little to do with the old-fashioned idea of "development" he was familiar with and often espoused. He might have been surprised by our current fears about global warming, and by some predictions about its probable effect on the Arctic. It would have been interesting to hear his ideas on the subject.

This book was researched and written in a very short period of time, with no grants, research assistants, or other academic underpinnings. I would like to thank my wife Marilyn Carson, and all my family and friends, for putting up with me during this time of concentrated effort. I would also like to thank my editor, Michael Carroll, for thinking of me as someone who might be willing and able to do this biography. I enjoyed living in Stefansson's world for a few months, and I hope this small book will induce a new generation to take note of his amazing adventures and rich and complex career.

◌

Note: Most of Stefansson's works and the works of his contemporaries use the designation "miles" to express distance, (still the practice, of course, in the United States), and the designation "Eskimo" when they speak of the Native people whom we currently know as "Inuit." It would have been extremely difficult, rather pedantic, and historically incorrect with respect to the texts, to attempt to change these terms where they appear, so as to conform to current Canadian practice. Therefore the reader may notice some discrepancies, since "Eskimo" and "miles" will appear in the older texts, while I use the designation "Inuit" when speaking about the present and the metric system where I have estimated distances myself.

Tom Henighan

Stefansson at eight years old, with his sister, Rosa.

1

A First Glimpse of "Stef"

June 13, 1919
The Harvard Club
27 West 44th Street
New York, N.Y.

A letter from Bronson Hardwick, Harvard scholar and researcher, to Charles Alcott, a former classmate, describing his first acquaintance with the explorer Stefansson:

Dear Charles,

Well, here I am in the great Gotham itself, ensconced in the gloomy old Harvard Club on 44th Street, in a big dark room, already littered with my manuscripts, pre-war travel notes from

Egypt, and piles of those dull Loeb Classical library volumes that are proving so helpful in the more obvious parts of my research.

You remember, I told you about my idea of writing a book on the Greek historian Herodotus? It's a project I've been promising myself — and some not over-eager publishers — ever since I first started following the routes of that grand old traveller and reporter through the eastern Mediterranean. (I call him a "reporter" because that's what he was, in addition, of course, to being one of the first classical historians, and the one most fun to read). Well, I thought I would move in here at the Harvard Club — the rooms are cheap! — use the various libraries in the area by day, get this project done, and still find some enjoyment in the metropolis by night.

Now, having settled in, and not having seen you for some months, it struck me as appropriate to write, if only to elicit one of your famous droll letters. Really, it's so long since we were undergraduates together! This European war has made the years pass very slowly; one felt so cooped up, not being able to visit over there, but now that the fighting's really done, we can get on with the fun and let dour old Wilson go about making the world safe for democracy. A good time, too, to get down to a favourite project, one too long postponed, or so it seemed to me. But more on that later!

First, I've got something much more interesting to share with you!

So far I haven't seen much of the city, but thanks to living in these gloomy quarters I've made a new acquaintance — he has "digs" just down the hall — and I'm eager to tell you about him. You may even have heard of him if you've been reading the papers pretty carefully over the past few years — he's done a few amazing things and everything about him is striking, starting with his name, which is Vilhjalmur Stefansson! Quite a mouthful! That first bit sounds something like Vil-YAL-mur, but he likes to be called "Stef" — he's pretty easy-going and completely unpretentious.

Although we'd exchanged a few dutiful nods in the hall, I didn't actually speak to him until about a week ago, and the way it happened was somehow appropriate. I had got up early as usual and was trotting out to breakfast, with library duty to follow, when I saw our grizzled old iceman with his rough wagon and decrepit nag stopped on the street. The old geezer had picked up a big block of ice in his tongs and was groaning and swaying down the alley to the kitchen, ignored by all and sundry passers-by. Only one man, who stood nearby, had stopped to watch. I recognized him immediately as Stefansson, my club neighbour down the hall, although at that moment the explorer seemed rather transformed.

I should explain that Stef is a strong-bodied sharp-faced, European-looking fellow somewhat above average height, trim and wiry, who nonetheless seems to take up a lot space, one who looks tidy enough to be a town swell yet sturdy enough to be a welterweight boxer. He's a slightly more refined version of one of those Neanderthals the anthropology chaps dress up in modern clothes to show how much they look like us. At that moment he was gazing intently at the iceman, or rather at the man's gleaming burden. To my amazement, as the old geezer staggered past him, Stefansson reached out quickly and touched the block of ice. Then his hand came quickly up to his face, he sniffed at his wet fingertips, smiling, and with eyes half closed, tasted the ice, almost as if he were evaluating it, or conjuring up some scene from long ago. It was exactly the way I've seen some of our old New England farmers taste their soil to test it. I stood amazed. What struck me was that this rather dapper fellow, dressed in city fashion with suit and fedora, seemed to have been transformed into something either pre- or ultra-urbane, into a primitive or an aesthete, or more accurately both at once. I swear to you that at that moment the man seemed directly in touch with his sensory experience, unselfconscious, perfectly natural. His little gesture was purely spontaneous, yet at another level, it appeared fully calculated, as if he were savouring and evaluating the ice and his own relationship, as explorer and showman, to it. I don't think he looked closely at the iceman at all.

Shortly afterward, we were walking side by side down 44th Street. He had observed me observing him, and had immediately approached me and suggested that we have some breakfast together. I accepted with delight; I had never met an explorer before, and found his sturdy presence, direct manner, and hearty geniality appealing. This was a man to whom it would be hard to say "no."

So I had a few hours to observe him and listen to him, and based on that and a few further meetings, I have a sharp sense of the man, and hopefully a quite accurate one.

I found him eccentric but charming — and a man who seemed comfortable declaring his convictions. These never came across as dogmas; he almost always referred his notions to his experience. At that first breakfast of ours he ate only bacon, ham, and some fish cakes, refusing all bread products. He explained to me that carbohydrates, in his opinion, were quite unnecessary, and probably even detrimental, to human health. He had shared the Eskimo diet for many years, and also studied it very closely, he told me, and had drawn certain conclusions about the dubious path of human eating habits when certain kinds of food were overused. His mother, he told me, had grown up in a village in Iceland and ate almost exclusively milk products, meat and fish, and she had never heard of a case of tooth decay among her people. He also alluded to historical examples — including, to my amazement, the Roman — for he turns out to be quite a good Latinist — and talked about the fatty meats, fish, and healthy wheaten gruel many Romans ate. The delicate luxuries of Trimalchio's feast that Petronius wrote about in his famous Roman novel, Stef insisted — quite rightly, I think — were not at all typical of the Roman diet. Stef himself, when he was in the North, made sure there was plenty of fat in his diet, and these days he is careful to stay away from carbohydrates, and even vegetables. He attributes his fine physical condition and good health to this choice.

We've met quite a few times since that first day, and I've learned a lot about his ideas and listened with fascination as he recounted some of his experiences. For one thing, I soon saw that Stef has a great feeling for animals. While we sat at table in a nearby café, a woman crossing the street and visible through the window fussed and stroked at her dog, a lively terrier. My new friend shook his head and tut-tutted. "That poor dog is doomed to a hard life," he insisted. "Despite her demonstrative affection, his mistress has no empathy at all with him, and the dog knows it. You can see it at a glance ... You'll notice the way she touches him and pulls the leash at the same time. " He paused and smiled: "An attractive woman, though, one must admit." That was only the first of many times that I noticed how sympathetic he was to animals — dogs in particular — and what a connoisseur he was of women who were beautiful and chic.

I hope I am not making him seem opinionated or pompous, or irritatingly self-conscious. He is the most natural and the least pompous man I have ever met. You would know that almost by looking at him. As I've said, he has a rough-hewn appearance, but is very compact and tidy. Perhaps it's his thick and sometimes unruly blond hair, which is streaked with grey — he's in his forties, after all — which makes him seem a little Viking-like. He has a well-shaped broad face, large sensual lips, an imperial nose, and high cheekbones — a very Scandinavian look — and his sharply defined cleft chin speaks of stubbornness and strength. His blue eyes are alive and striking and often full of irony and amusement.

Charles, I wish you could hear him speak about exploring, adventure, and the North, which he does in his fluent way, but without exaggeration or self-inflation, and with a focus on the practicalities of leading an expedition. It's hard to believe that he started out in university wanting to be a poet, but, as he puts it, he now sees himself as a "poet of deeds." The adventurers he admires most are the ancient voyager Pytheas, who lived around the time of Aristotle and is supposed to have sailed to England, Norway, and maybe even Iceland; Erik the Red; and Fridtjof Nansen, whom we all know for his great free-floating voyage through the polar regions in the 1880s, and of course for his noble efforts on behalf of world peace since then.

Stef is very well-informed about the past, and knows all the present heroes of travel. When I mentioned Robert Peary and his claim to have discovered the North Pole in 1909 (disputed of course by Frederick Cook), Stefansson grew wary. Although he spoke heartily in praise of Peary, and supported his claims, he seemed unwilling to go into details about Peary's personality or travels, and I had the feeling that Stefansson sensed that the Peary–Cook controversy, and other aspects of Peary's adventures, constituted a kind of bottomless pit which he was reluctant to be drawn into.

Stefansson seemed quite open about his own journeys and discoveries. He made three trips to the Arctic, he told me; the last one, just completed, lasted five years, which to me seems an inordinate amount of time to spend in the northern regions — or in any primitive spot! On his first trip, which he began shortly after he finished his studies at our own Harvard's Peabody Institute (that's how he gets to reside at the Club!), Stef met a wild character, a sea captain named "Charlie" Klengenberg, a regular Wolf Larsen outlaw type, right out of a Jack London novel. Klengenberg had stolen a schooner, plundered a storehouse, and sailed off with a bunch of cronies to carouse a little on god-forsaken Victoria Island. (If you draw a line from Greenland toward Alaska on the Mercator map you'll see just how isolated and how far north that huge island is!) Anyway, Klengenberg ran afoul of his crew, murdered a few of them, and coerced the survivors into silence. Stef met this notorious character on Herschel Island (it's near the mouth of the Mackenzie River to the west of Victoria Island), and this horrible man told the explorer that he had spotted some very European-looking Natives on Victoria, Eskimos with blue eyes and blond hair no less! Other people Stef talked to doubted this report — after all, there had been a lot of boozing going on among Klengenberg's gang — but Stef was intrigued. His Icelandic background and his interest in the Vikings led him to recall that the Greenland Colony founded by Erik the Red had mysteriously disappeared a few centuries after the year 1000 A.D. All sorts of reasons had been advanced for the demise of the colony, but a pretty cogent one seemed to be the possibility of intermarriage between Natives and settlers, which might have led eventually to a trace of Viking blood in the Native population. Or so Stef surmised. Could it be that the scoundrel Klengenberg had really spotted some descendants of the original North

American Vikings on remote Victoria Island? Stefansson was determined to find out.

That was the main focus, he told me, of his second expedition, and in 1910, near Victoria Island, he did finally encounter the legendary "blond Eskimos." They were actually an isolated and reputedly fairly xenophobic tribe also known as Copper Eskimos, because they made use of that metal. The resemblance of many of these people to Europeans proved to be very real, and even now remains something of a mystery, despite much speculation. (You and I Charles, I assume, must have first heard of Stefansson when the newspaper reports of "lost Vikings" began to appear.) Well, I had breakfast with the man himself the other morning in Romany Marie's curious restaurant in Greenwich Village. (It's a haunt of artists, writers and scholars, although goodness knows why, since she doesn't serve any hard liquor, even at night, but Marie herself is a great and famous character.) There we sat, at the table she usually reserved for him, and Stef told me that, in the words of the poet Byron, he had "awoken one morning to find himself famous." Then he elaborated a bit: "Those newspaper fellows, as you know, are usually more concerned with a good story than they are with factual reporting of scientific discoveries. Although there's no evidence against the Copper Eskimos being connected with the Vikings, there's no evidence for it either. The question remains open. The newspapers, however, chose to headline the supposed connection, and even when I gently corrected some of their exaggerations they were pleased to ignore my strong qualifications. The result was that many of my colleagues remained skeptical about the very existence of these Eskimos with European features that I had brought to the attention of the world. For me, the sad result was that I was regarded in some quarters as a romanticist, a publicity seeker, and wild speculator rather than as a good and careful observer. And when I think of all the exact observations I wrote in my notebooks, and how carefully I chose my words when I talked to the press, so as not to seem to exaggerate the similarities between the Copper Eskimos, Europeans in general, and Scandinavians in particular!"

Stef smiled, sipped his coffee, and waved to a very pretty woman who sat a few tables away. "An old friend," he explained, "a magazine editor. Sooner or later everyone turns up here. Maxwell Bodenheim, the poet, the playwright O'Neill, who will do great things if he doesn't drink himself to death, the novelist Willa Cather, and many others — they all show up here eventually. You may wonder what interest all this has for an explorer like me, for someone who has spent so much time hunting polar bears, eating raw fish, and living in snow houses. But my first ambition was to be a writer, and I intend to spend whatever years I have left on earth writing about things that interest me and which I've investigated — the life of the Eskimo, the Arctic environment, diet and health, the history of exploration. I'm hiring a researcher and a secretary and I've begun to plague the magazines with my articles. So perhaps I'm not out of place here after all."

Stef was being quite modest about his the range of his interests, as I can testify after my various conversations with him. He has mastered various Eskimo languages and dialects and had fascinating comparisons to make between them and English; he had shrewd comments on all aspects of northern culture — some of his stories about the awkward grafting of Christianity onto the shamanistic Eskimo beliefs were worthy of a Sir James Frazer. He knows a great deal about hunting, fishing, and about survival in an extreme climate, but is also well-versed in history and literature. In general he is a very well-read man, sensitive to the arts, yet quite capable of walking through Central Park with an eye alert for wind patterns, the flight of birds, and traces of the Ice Ages. His Icelandic background, Canadian birth and connections, and American education have given him some unique perspectives, while his various journeys and encounters in the northern wilderness have deepened his vision of life, so that he often seems something of a seer or a prophet, a role he might discount with a healthy burst of laughter.

When I talk to him I find it difficult to believe that he won't soon be off again to some remote, harsh region, the Far North, the South Pole, God knows where. He is, as I've indicated, very much the rational fellow, and thinks that intelligence, good sense, and empathy with the Natives and the environment can override almost all dangers. But I have a feeling that he may be destined for a few trials greater than any he's experienced so far. Or quite likely, he hasn't told me everything of his past troubles. There's a kind of blithe confidence, a hubris about him that the gods, Norse or otherwise, may very well want to chasten.

We should both keep an eye out for his name in the newspapers.

That's all for now, dear Charles, but perhaps in some future letters I will have more to tell of my adventures in New York, and more detailed accounts of my meetings with the redoubtable Stef.

Your old friend from our never-to-be-forgotten "Crimson" days,

Bronson Hardwick

Students at Grand Forks, circa 1897. Stefansson is standing, third from the right.

2

Childhood, Youth, and College Days

Vilhjalmur Stefansson was born in 1879 in the small Icelandic settlement of Arnes, near Winnipeg, Manitoba. His family had immigrated to Canada under a Canadian government recruitment program that sought skilled agricultural workers to cultivate the prairies. Sadly, this particular project was doomed to failure. The farming land offered was poor, and bad weather, malnutrition, and disease soon decimated the little colony. Two of the Stefansson children died, and the elder Stefansson decided to relocate south of the border, in the Red River Valley in North Dakota, where a more favourable site had been pioneered by earlier Icelandic emigrants.

Here the young Vilhjalmur grew up, speaking Icelandic at home, learning English with his parents and in his intermittent primary school experiences (which totalled only a couple of years), and relying for most of his education on his father's example of voracious reading.

The boy Stefansson was fortunate in that the Icelandic tradition encouraged love of learning; among his first reading at home were the Norse sagas, and the Bible (the family was Lutheran). The sagas, no doubt, imbued Stefansson not only with a love of adventure, but with the sense that life was a challenge that could best be met by cultivating both physical and mental abilities, while the Biblical narratives offered a spectrum of mythologies and moralities that must have partially stimulated the young man's later interest in tribal lore, customs and mores, belief systems, and behaviour.

Young Stefansson's experience and early upbringing on the frontier was individual enough, but hardly unique. Readers of Willa Cather's fiction and essays will recall her focus on those young men and women who became the protagonists of her novels and stories, frontier heroes and heroines who achieved success after emerging from one or more of the immigrant groups that she had encountered during her own youth in Nebraska. The children of Swedish, Bohemian, German, French, and other settlers, as Cather observed, were very likely to be challenged and tested in ways that the native American stock was not; yet quite a few of the former ended up as doctors, lawyers, professors, or writers, surpassing most of their native-born contemporaries, despite childhoods that were underprivileged in terms of conventional middle-class benefits.

Stefansson's family broke up when his father died in 1892. The boy was only thirteen, and spent time with his older brother Joe herding cattle, sleeping out on the range, and learning to hunt antelope. A few times, in the fierce Dakota winter storms, he came close to freezing to death. Although he had no contact with well-educated people, his intellectual curiosity, thirst for knowledge, and determined reading, saw him through. At age eighteen, though nearly penniless, he entered the University of North Dakota, severely handicapped though he was by his lack of formal preparation, a defect which he had to overcome by taking many special exams and courses.

Stefansson in college was a living exemplar of the old American dream of winning success through hard work and determination. He took courses, worked in the hayfields, taught younger students, and jumped at every opportunity to get ahead, taking on everything from milking cows to assisting in the biology lab. "He got his room free for assuming the responsibility for ringing the study bell at the boys' dormitory. An agent for a city steam laundry, he got fresh washing. All the money he needed to raise was ten dollars a month for his board." Yet he was far from a docile young man, and was thrown out by one shocked landlady who heard him praising Darwin and criticizing the Lutheran Church (Hunt 6–7).

Stefansson was an outstanding student, but a lively, and even an obstreperous one. He constantly challenged the faculty, sometimes mocked them, and was generally a leader in the areas that interested him — debating, poetry, and Scandinavian studies, among others. At last, however, his absences from class and a prank played on an instructor in a German class led the faculty to suspend him. The students were aroused and threatened to strike in protest, but he dissuaded them; instead they staged a mock funeral, trundling him off in an improvised wheelbarrow-hearse, followed by a girl in widow's weeds, and a long procession of his friends. His suspension raised some eyebrows in provincial North Dakota, and he was offered a job on a newspaper in Grand Forks and other opportunities. He decided, however, to continue his quest for a college degree elsewhere. He wrote to several universities, asking if they would give him credit for passing exams even when he had not taken the requisite courses; a few agreed to this, and from among these Stefansson chose the University of Iowa as his next academic stop.

He arrived in Iowa City early in 1902, and soon plunged into serious academic work, taking such subjects as Old Norse, Norwegian, German, Spanish, Swedish, Latin, and Old English. He obtained the B.A. degree a little over one year later, an astonishing achievement, which, as his biographer Earl Hanson notes, was cited by an Iowa University psychologist as evidence of what could be accomplished by accommodating especially gifted students.

From Iowa, Stefansson went east. Through the intervention of two leaders of the Unitarian Church from Winnipeg, he was offered a scholarship to study at the Harvard Divinity School. After a year there, however, he decided that he was on the wrong path, and, recognizing his own broad interest in so-called primitive societies, their languages, customs, social structures, and adaptation to the environment, he turned to the study of anthropology. He obtained a scholarship at the Peabody Museum, and found a mentor in its director, Frederick Ward Putnam, who was also head of anthropology. Putnam suggested that Stefansson focus on African studies and he was urged to join a 1906 British expedition to Central East Africa. He had already plunged into serious preliminary study of Africa when a fledgling Arctic expedition in need of an anthropologist applied to the Peabody. Stefansson was recommended. This led the young man to a turning point.

Up to that time, Stefansson had very limited foreign travel experience. However, the trips he did take were significant, for he had gone briefly to Scandinavia, and on two occasions had visited Iceland, his ancestral country. On one of his Iceland trips, in 1905, he collected skulls from an old cemetery, and upon studying the teeth he noticed the absence of signs of tooth decay. He recalled that his mother had claimed that tooth decay was extremely rare in rural Iceland, where the diet was quite different from that of North America. Stefansson's life-long interest in nutrition and diet and their physical consequences had its origin at this time. Both his Iceland trips and his study of traditional dietary rules and taboos in his courses in religious history at Harvard were strong factors in promoting it. The Iceland trips no doubt also drew Stefansson back to his roots in northern culture. He was well aware of the connections between the Norse settlements there and the later Greenland adventure of Erik the Red. The invitation to travel north must have seemed like a summons from destiny: here was a wide open sphere of field study for a young anthropologist like Stefansson, and one that might be all the richer in the light of his own deepest cultural affinities.

About this time, Stefansson met the person who was probably his first great love, Orpha Cecil Smith, a Boston student several years younger than he, who had come from Toronto to study drama. They became engaged, probably in the spring of 1906, but when Stefansson ventured north, he enthusiastically accepted a lifestyle and career that pushed the lovers apart. They stayed in touch for a while, but in 1910 Orpha Smith married an old friend, an inventor named Harold Sheibe, and her last meeting with Stefansson took place in New York in the 1920s.

Snowhouse complete with entrance passage and sled about to be unloaded. Stefansson-Anderson Arctic Expedition, 1908–1912.

3

First Trip to the North, 1906-1907

Stefansson's first field trip, the "Anglo-American Polar Expedition" was hatched by two young men, Ejnar Mikkelsen, a Danish naval adventurer, and the American geologist Ernest de Koven Leffingwell. Both had been to the Arctic a few years before, on a venture that was less than satisfactory, and they decided to join forces in what they hoped would be a more personally rewarding one, both scientifically and in terms of renown and profit. They had gone about gathering funds both in England and the United States, and, despite some setbacks, were able to proceed with grants from the Royal Geographical Society and such private benefactors as John D. Rockefeller. They sailed north in 1906 from Victoria, British Columbia, to Herschel Island, their chief goal being to explore the Beaufort Sea and the continental shelf north of Alaska.

Stefansson, who agreed to join their expedition on his own terms, decided to make his own way north, so as to become acquainted along the way with the lands and Native peoples that interested him. He travelled via Edmonton to Athabasca Landing, took a steamer north to Grand Rapids Island, and from there sailed on a stern-wheeler to Fort McMurray. He went on to Athabasca Lake and the Slave River, crossed the Great Slave Lake in a steamer and reached the Mackenzie River. An Inuit whaleboat carried him on the long journey up the great river to Herschel Island.

Tiny Herschel Island lies in the Beaufort Sea, about one hundred kilometres east of the extreme north-eastern border of Alaska. To the east of Herschel, the coast juts out forming Cape Dalhousie and Baillie Island. A few hundred kilometres east and north, on the Amundsen Gulf, lie two very substantial land masses, locatable on every northern map, Banks Island and Victoria Island, the latter of extreme importance in Stefansson's career.

Stefansson's first trip north was made in the summer of 1906, and a few vivid impressions remained with him the rest of his life. Clouds of mosquitoes plagued him — he had never encountered the like — and on some days the temperature climbed over 100°F. Superb vistas unfolded: the Great Slave Lake, where they sailed out of sight of land, seemingly endless forests, complex river systems, great mountain ranges, and the vast northern sea. The wide reach of the Hudson's Bay Company's enterprises impressed him, and he took note of the mutually dependent system of connection between the Native hunters and the trading post operatives. Above all, however, Stefansson was intrigued by the Inuit, and drew conclusions from his experiences that would shape his future life and career. It was very characteristic of him to form quick impressions: almost at first glance, he saw the Inuit as a fine sturdy people, whose character and traditions had been, on the whole, only minimally affected by their contact with the white entrepreneurs from the south.

As expedition anthropologist, Stefansson's assignment was to study the Native peoples of the Mackenzie River area and to collect artifacts for the Peabody and the Royal Ontario Museum. Stefansson, however, anticipated with even more excitement the contact he would have with the Inuit people. In fact, the Inuit were experiencing some changes of lifestyle in 1906, since southern foodstuffs, which they had in recent years obtained from white traders, were not just then available. As a result, they had reverted to many traditional practices. Sailing up the Mackenzie River in a whaleboat with an Inuit named Roxy, and afterward, when he ventured east from Herschel Island, Stefansson got rid of some preconceptions. He found that while the Hudson's Bay Company's discriminatory behaviour toward the Mackenzie River Natives had suppressed them, the Inuit, still able to draw upon their traditional practices, were more independent. Also, contrary to what he had imagined, the Inuit were in general not at all "short" people. Nor was their saltless fish diet unacceptable to any white man who was willing to give it a fair trial. As for housing and apparel, no southerner had any reason to shun their traditional dwellings, while their clothing and transport — geared for survival in every way —should be, he decided, the basis for all sensible Arctic land travel.

When Stefansson sailed to Herschel Island with Roxy he had his first introduction to the ice floes of the Arctic Ocean. These were one of the traditional terrors of Arctic travel, but Stefansson, reassured by the Inuit, saw that they had some benefits: for one thing, the ice blocks and the piled up mounds of ice tended to create smooth patches, known as "leads," in the often rough Arctic waters; for another, they provided a ready supply of fresh water, since the ocean salt, over a period of time, works down through the ice, leaving the surface ice potable.

At Herschel Island, Stefansson found a unique community, one that reminded him a bit of the atmosphere in one of Jack London's famous novels, *The Sea Wolf*. A small island — only thirteen kilometres long and about six and a half kilometres wide — its sheltered harbour had made it a favourite base camp and stopping place for whalers, fishermen, explorers, and other non-Natives with business in the region.

It was on Herschel Island that Stefansson met Christian "Charlie" Klengenberg, an outlaw who had stolen a ship, plundered a storehouse for provisions, and sailed away to Victoria Island in the east. There, in the course of some drunken carousing, Klengenberg murdered his chief engineer, and on the way back to Herschel, effectively silenced three more crew members by dispatching them as well. When Klengenberg reported to the RCMP that the missing men had suffered accidental deaths, his crew corroborated his story. But when he fled from the island they implicated him, explaining their false statements on the grounds that he had threatened to kill them if they told the truth. Stefansson, who took some of the testimony of the sailors, also heard from them a strange story about the Native inhabitants of Victoria Island, whom, they reported, were Inuit with distinctly European features, including, in some cases, blue eyes and blond hair.

Stefansson had intended to visit Victoria Island, but this report — although discounted by the experienced whalers on Herschel — turned his intention into an imperative. Here was a real treasure for an apprentice anthropologist. And to Stefansson it had a special significance, since he immediately speculated that there might well be some evidence that members of Erik the Red's medieval Norse colony in Greenland, which had disappeared without explanation by the fourteenth century, had intermarried with the Inuit and passed some of their physical characteristics to these isolated island people.

Autumn was coming on and there was still no news of the ships carrying Leffingwell and Mikkelsen northward. Stefansson decided that he was wasting his time on Herschel Island; if he wished to become more closely acquainted with Inuit culture he must move on. He sailed eastward with Alfred H. Harrison, a British traveller, and reached Shingle Point, eighty kilometres east, where he remained a few months, living with the Inuit.

Stefansson learned much from his hosts. He hunted seal and polar bear with them, and under their tutelage he became an expert dog-team driver. He became familiar with the local environment, and noted how the Natives made use of it to build their rather uncomplicated but strong culture. Each item of Inuit clothing, each piece of equipment they used, and the shelters they inhabited — all attracted his meticulous attention. Not only did he record this data in his notebooks, but he also mastered their hunting techniques and learned how to use their tools (Hunt 24).

Excellent and enthusiastic linguist that he was, Stefansson also set about learning the Inuit language and dialects. In October 1906 he moved still farther east, to Tuktoyaktuk, some 160 kilometres from Shingle Point. He was drawn there by the presence of an Inuk named Ovayuak, who not only lived in the traditional Inuit manner, but also spoke various dialects. Here Stefansson began his systematic study of the Inuit language and began to master the art of survival in the traditional Native style. As Stefansson's early biographer Earl Hanson tells it:

> The house at Tuktoyaktuk was made of the usual framework of driftwood, the roof supported

by wooden pillars. The middle area was the general living room; the recesses were used for sleeping. The earthen walls were six feet thick at the base, sloping upward on all sides to a flat earth-covered roof, in the middle of which was a skylight made of sewn strips of polar bear intestines. Light and heat were furnished by four soapstone lamps, each shaped like the letter *D*. The bowl was filled with oil rendered from seal or polar bear blubber; and along the straight edge of the *D*, for a wick, was strung a ridge of powdered wood or walrus ivory. Sometimes dried moss was used instead. A sheet-iron stove was used for cooking, and when it was lit the temperature often went up to one hundred degrees (Hanson 44).

Hanson's account makes evident how at every turn Stefansson must have observed the Inuit's dependence on their environment for their basic needs and traditional cultural artifacts. But he not only observed; he participated daily in a routine that included few frills, one sustained by long-practised arts of survival in a difficult climate. Hanson describes the routine:

He was usually aroused in the morning by somebody striking a match to draw an early puff or two on a pipe. This awakened others, who followed suit, conversing across the recumbent forms of their neighbours. Then some of the women, dressing hurriedly, would run out to the alleyway [the space between the shelters], returning with armfuls of frozen fish, which were thrown on the floor to thaw. When the fish had slightly thawed, the women would take their half-moon shaped knives, cut off the heads (to be saved for a later, cooked meal) and make a straight slit from neck to tail along the back and the belly. Then, catching the skin in their teeth the women would peel a fish as one would a banana. The best pieces, chiefly the heads and tails, were saved for the children and the visitors, and the rest would then be placed on various trays, ready for eating.

Breakfast finished, all would disperse to their various occupations, which at Tuktoyaktuk consisted mainly of fishing ... (Hanson 45).

Beginning with such early experiences, and continuing through his later expeditions, Stefansson learned much about the social and domestic beliefs and customs of the Inuit, which he recorded for his audience of educated readers quite objectively, and sometimes with amused detachment. The traditional Inuit family life, for example, included some practices that, however suited to their life conditions, could hardly find favour with the missionaries. Not only did they occasionally, and under very specific conditions, exchange wives, but they would sometimes expose unwanted babies, especially girl babies, a practice anathema to the clergymen labouring to "Christianize" them. Among the Inuit, modesty with regard to the body was more common among men than women; in the close atmosphere of their dwellings, and sometimes elsewhere, the women thought little of uncovering their sexual organs, while the men were much more guarded. Pregnancy was often taken casually and childbirth seemingly unanticipated; little connection was made between the sex act and conception. Children remained a long time on their mother's back — for they were carried everywhere — and continued to feed at her breast for a few years. As to cleanliness, Stefansson — in contrast to other explorers — defended the Inuit, suggesting that despite the climate they enjoyed bathing, and that their "rank" dwellings seemed so only to visitors not yet used to the smell of raw fish. Ironically, he later criticized their frequent bathing practices — acquired after they were Christianized — because these involved various unhygienic practices, and so became a prime means of spreading germs.

Stefansson often adopts a tone of wry tolerance with regard to Inuit beliefs and shamanic practices, and gently mocks their selective and somewhat distorted way of adopting Christianity. Yet he never lost sight of the fact that, despite the tribalism they shared with many so-called "primitive" peoples, they had developed a remarkably peaceful and socially harmonious way of life. There were no Inuit wars, very little crime of any kind, and a type of benevolent communism that was probably the closest real equivalent on Earth to some of the dreams of the classic Utopian writers. Whether the Arctic was actually as "friendly" as Stefansson claimed, it is certain that the people who inhabited it were hospitable, non-violent, and unaggressive in ways that might well be envied by those living in certain sectors of modern industrialized society, both in Stefansson's time and ours.

In March 1907, Stefansson finally learned the whereabouts of his fellow explorers Leffingwell and Mikkelsen. Although their voyage had been a troubled one in many ways, ending with the loss of their main ship in the pack ice, they had subsequently travelled by dogsled far into the Arctic Ocean.

While they found no new land, they learned much about the ocean depths. (Leffingwell would later make important discoveries about permafrost.) Tentative efforts to prolong the expedition or to reconstitute it, with Stefansson included, failed, and he himself — after a few months of archaeological work — finally headed south. His notion of his life's work, however, had by then completely changed; from that time forward he was fully prepared to "follow his bliss" and to undertake a much deeper investigation of the Arctic region and the life of its indigenous people. As Earl Hanson puts it: "The North was in his blood now, and all dreams of working in Africa were gone. He had had a taste of life in the Arctic and he yearned to return to it. Day and night he thought of those Stone Age people on Victoria Island who were said to resemble white men, and the thought filled him with determination to organize his own expedition and to visit them."

The Copper Inuit. Coronation Gulf. Glass plate.

4

The Second Expedition: Meeting the "White Eskimos"

Settling down temporarily in New York's bohemian Greenwich Village, where he enjoyed the company of artists, poets, and eccentrics of various stripes, Stefansson soon began planning his second expedition. With the backing of the American Geographical Society he wrote popular articles on his first experiences of the North, eventually enlisting the support of the American Museum of Natural History and the Canadian Geological Survey for a trip to begin in the spring of 1908. Both the Museum and the Society were impressed by the young anthropologist's conviction that he could mount an expedition for minimum costs — made possible because he would live in the efficient manner of the Inuit. Also intriguing was the notion that he was on the track of isolated Inuit groups, the study of which might contribute to a wider and deeper knowledge of the cultural and racial connections of the northern peoples.

Although he had originally intended to travel alone, by "a stroke of luck" Stefansson was approached by Rudolph Martin Anderson, a friend and former classmate from the University of Iowa, who asked if he might join the expedition. As an undergraduate, Anderson had been Phi Beta Kappa, Sigma Xi, and an all-American athlete, and he later became a Ph.D. in ornithology and published a standard book, *The Birds of Iowa*. Stefansson never recanted on his initial profession of delight at connecting with Anderson, yet in 1908 he could hardly have suspected what an important role the man would play in his life and career. In fact, even after years of dealing with the resistances and negatives emanating from Anderson and others connected with him, Stefansson never seemed to grasp that he had become a kind of "shadow figure" for Anderson. By which I mean that Anderson was obsessed by him, just because he saw Stefansson as achieving what he himself wished to accomplish, but couldn't. Anderson's hatred was the enmity of a repressed man who despises his extroverted, successful, and confident friend for being what he could never be (but secretly, and unconsciously, wished to be). Despite Anderson's evolving hostility, the genial Stefansson, never fully "called him out." He gave Anderson full credit for his various achievements in their joint explorations. He preferred to ignore the negative side of their relationship, admitting that they had had their differences, but refusing to assume any malice on the part of his friend, co-explorer, and adversary. It was the same approach he would take in the similar but relatively trivial instance, when the famous explorer Amundsen heaped scorn on Stefansson's idea of "the friendly North."

The conflicting emotions that so obviously simmered beneath the outwardly civil relationship between Stefansson and Anderson never quite burst out in dramatic fashion as they had, for example, in the famous case of Richard Burton and John Manning Speke (who died, very possibly a suicide, shortly before a scheduled public confrontation with Burton), but that they were deeply present is clear from the account of one of Stefansson's biographers, D.M. LeBourdais.

When, in the late 1920s, LeBourdais attempted to interview Anderson in connection with his research on Stefansson, he was thrown out of the scientist's Ottawa office: As LeBourdais tells it, "Anderson rose to his feet, his face flushed. 'Mr. LeBourdais,' he shouted, as he came round the end of his desk. 'I did not invite you into my office and I ask you to get out!'" According to LeBourdais, "Anderson never made any effort to disguise his antipathy toward Stefansson, or displayed much discretion concerning those to whom he unburdened his feelings. The vendetta was his ruling passion, well known to his associates over a long period. That he was able to infect with his own antagonism other members of the Civil Service, to some of whom Stefansson was little more than a name, is a remarkable circumstance" (LeBourdais 176).

Nevertheless, in early 1908 the two colleagues, still in harmony, left together for the North, following the path of Stefansson's first journey, but arriving on Herschel Island only to find that the basic supplies they had hoped to pick up there had not arrived. Not only that, when they sought to mount a local expedition with Inuit and dogsleds they found themselves in a serio-comic situation, being unable to obtain the matches that were essential to the Inuit (and to Anderson), all of whom were addicted smokers, and would not travel without them. Stefansson ended up going west, and near Point Barrow, Alaska, finally located the whaler *Narwhal* with the supplies sent up by the Museum of Natural History. Unable to return at once to Herschel Island, he remained in Alaska and did anthropological research there.

The following summer, while Anderson worked in Alaska, Stefansson travelled east on the whaling ship *Karluk* (which was to be a fatal ship in his later career) and landed at Cape Parry, a rocky peninsula south of Banks Island, some four hundred kilometres east of Herschel Island. With him was an Inuit man named Natkusiak and a seamstress named Pannigabluk, who was to become Stefansson's bed companion, and the mother of his son.

Difficult months followed. Anderson did not immediately join the party, and Stefansson travelled west to find him. They then moved east again and Anderson stayed at Cape Parry while Stefansson hunted inland; winter came on and food supplies were scarce. Eating only lean caribou meat — and not much of that — Stefansson and his party got sick, which subsequently restricted their hunting. Anderson, too, got sick, and Stefansson returned to Cape Parry to make sure of his recovery. Stefansson remained there from January to March 1910, at which time Anderson, then fully recovered, headed back to the Mackenzie Delta to pick up mail and supplies from the south.

It had been a frustrating start to their expedition, but at last, on April 21, 1910, Stefansson set off for Victoria Island, in search of the rumoured tribe of "White Eskimos." Travelling east by land and accompanied by Pannigabluk, Natkusiak, and Tannaumirk, Stefansson and his party garnered an adequate supply of both seal meat and caribou. Through the difficult winter he had learned much about survival, yet "these mysterious people" that he sought seemed as far away as ever. The journey continued until they were forced by a blizzard to halt and make camp, but the game held out. As Earl Hanson tells it:

They came near Dolphin and Union Strait, where Victoria Island comes close to the mainland. Here Richardson had explored almost a hundred years before; he had seen no Eskimos and that region had since been considered uninhabited. But suddenly, on May 9, just nineteen days after leaving their base at Langton Bay, Stefansson and his companions made the great discovery that there were human beings nearby after all.

It was merely a fresh adze mark on a piece of driftwood, as though somebody had tested the wood for soundness, but Robinson Crusoe, discovering a footprint on his island, could not have been more agitated…. That same day, at Cape Bexley, they discovered a deserted village of over fifty snow houses. Its size amazed them …

A broad trail led northward from the village across the sea ice toward Victoria Island (Hanson 71–72).

On the ice of Dolphin and Union Strait, the explorers finally located a remote but busily active Inuit village, where — after some doubtful preliminaries — they were welcomed. During their stay there, they were told about another settlement, on Victoria Island, a short trek north, where the men had eyes and hair that much resembled Stefansson's own. Eagerly, he set out to find what he hoped were the near-legendary Inuit he had come so far to meet. And indeed, some twenty-six kilometres away, on the island shore, Stefansson at last came face to face with the "White Eskimos" that were to constitute his major scientific "discovery" and to make his name as a celebrity in the newspapers back home.

Stefansson had in fact come upon the so-called Copper Eskimos. As William R. Hunt describes it, they "used copper knives ... fashioned from local ore, and from this metal, the only one they had, their name was derived. They cooked and heated their dwellings with seal-oil lamps. Unlike the western Inuit, who lived in half-buried huts of sod and driftwood, the Copper Inuit lived in tents during summer and in temporary snow houses which they built as they moved about in the winter. Their hunting weapons consisted of bows and arrows and spears, and their dogsleds provided them with effective transport" (Hunt 48).

The Copper Inuit seemed to Stefansson to have definite white characteristics: "There are three men here whose beards are almost the colour of mine, and who look like typical Scandinavians ... No one could fail to be struck by the European appearance of these people."

Diamond Jenness, an anthropologist who studied the Copper Inuit, cast doubt very early on Stefansson's racial connection theory; he attributed the Copper Inuit's variant physical characteristics to natural causes. In 2002, however, Professor Patricia Sutherland, in a lecture at the Canadian Museum of Civilization, summarized some recent archaeological work, which demonstrates convincingly that there were many contacts between the medieval Greenland Norse and the Dorset people of Baffin Island. In the light of this, the anthropologist Gísli Pálsson has commented: "There is every reason to wipe the dust off Stefansson's theory on the interbreeding of Norse and Inuit during the Middle Ages in the light of new methods, new knowledge, and new attitudes." Yet, as he points out, the contemporary Genetic History Project (of which he is part), while it does document contact between Norse men and Inuit women in Greenland, does not support Stefansson's specific attribution of possible Norse ancestry among the Copper Inuit.

Stefansson's mostly cautious speculation was not taken very seriously in his time, partially because it reached the public consciousness through various wildly exaggerated newspaper accounts, which made the explorer famous, but also — at least temporarily — undermined his credibility with the more conservative scientific community. References to his finding a "lost tribe of 1,000 white people, who are believed to be direct descendants from the followers of Leif Eriksson"; headlines that declaimed: "New Race Solves Mystery of the Ages," were certain to arouse the resistance of a skeptical scientific community. To their credit, many individuals, and news organizations including the *New York Times*, sought clarification, and allowed Stefansson space to present his views as they were, without fantastic embroidering.

Unfortunately, in the course of dealing with one controversy, Stefansson ignited a second, with some incautious remarks about the often negative influence of missionaries among the Inuit. "How dare Stefansson question the values of our Christian culture?" barked the *Boston Transcript* in a 1913 editorial. "He seems to be saying, 'savagery is health and vigor; civilization, with all its blessings, means decay.' And this coming from a former student in the Harvard Divinity School ... is indeed astonishing" (Hunt 62–63).

But if any proof was needed that Stefansson was hardly over-eager to "exploit" his newfound fame, or notoriety, it lies in the fact that he remained two more years in the Arctic after his first encounter with the Copper Inuit. During that time he rejoined Anderson and they travelled to Coronation Gulf, nearly suffering a fatal accident on the way due to carbon monoxide poisoning. While Anderson and others collected zoological specimens at Coronation Gulf, Stefansson, with Natusiak, crossed the Wollaston Peninsula on Victoria Island to Prince Albert Sound, meeting more Copper Inuit en route. During the rest of 1911–1912, Stefansson explored, made further study of the Inuit language, and did archaeological work between Cape Parry and Coronation Gulf, on the mainland southeast from Victoria Island. In late summer, 1912, having agreed with Anderson that they had accomplished enough to justify the expedition, he parted with his Arctic companions and returned to New York, seeking the bright lights of urban life and some months of longed-for intellectual stimulation, only to find himself embroiled in the "White Eskimo" controversy. Yet while the negative sides of his new-found fame were quickly brought home to him, his achievements also made it possible for him to plan his next venture. As Hanson tells us:

> Stefansson now wanted to organize an expedition that would devote itself in part to the exploration of the Beaufort Sea, and in part to general scientific research in many other fields. He also planned to go out on the ice, as he had always travelled on land, taking with him only a few supplies and living by hunting as he went. He had never been far out on the floating ice, but he was certain that it would be possible for a skilled hunter to go almost anywhere there, and find plenty of seals, polar bears, foxes, and

fish, which would provide him with food, skins for clothing, and fuel for heat and cooking (Hanson 104).

This was the beginning of Stefansson's third and last major venture to the North, the Canadian Arctic Expedition of 1913–1918, one that achieved much in the realm of geographic knowledge, but which was fraught with peril, and marked by a major tragedy that has threatened to tarnish the great explorer's reputation, even to the present day.

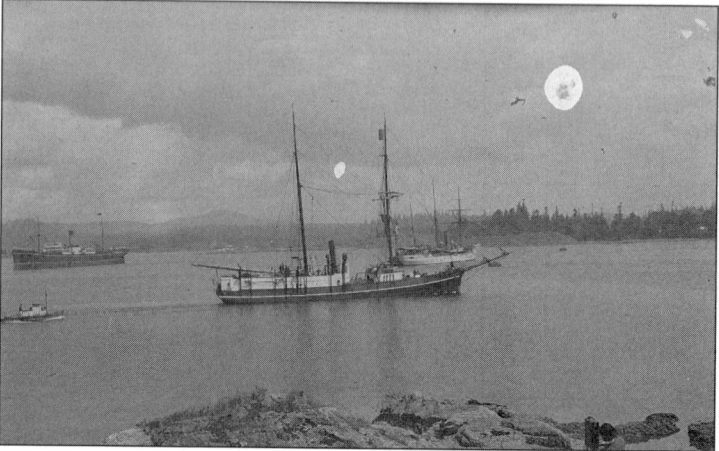

HMCS Karluk, *Esquimault, steaming about the harbour.*

5

The Canadian Arctic Expedition: Five Years in the North

Flush with his recent success, Stefansson began his quest for new funding by approaching the American Museum of Natural History in New York, but although they expressed interest, they planned to postpone their commitment, and he therefore turned to the National Geographic Society in Washington. That institution quickly pledged $22,500, easily sufficient to launch the project. Back to the Museum went Stefansson, and found them now willing to match the Society's pledge.

Meanwhile, he had invited Rudolph Anderson, then newly married, to join him as second in command, and acquired the services of Captain Theodore Pedersen, an old and skilled hand at northern navigation, who eventually chose the ship *Karluk* as the best available under the circumstances. Soon Canadian interest in the project materialized, and after some preliminary talk with the Royal Ontario Museum and the Canadian Bank of

Commerce, Stefansson was routed to the prime minister, R.L. (later Sir Robert) Borden.

The Conservative government of Borden almost immediately saw that the expedition promised important results, and would be carried out almost exclusively in northern Canadian territories or claimed jurisdictions. They agreed to take over the whole of the financing, provided the other sponsors could be persuaded to withdraw. The American groups agreed, and Stefansson received Canadian government backing, including agreement that he would be given "full responsibility and have the choice of the men going on the expedition and of the ship, provisions and outfit needed for the trip" (LeBourdais 65). The government official directly responsible for the expedition was G.J. Desbarats, Deputy Minister of the Naval Service, who little suspected what a tangled web of egos and ambitions he was to preside over during the next few years.

The expedition's chain of command, in fact, was set up in a rather complicated and awkward fashion. Stefansson himself, the designated leader, had worked to get Rudolph Anderson appointed to the Canadian Geological Survey, headed at that time by R.W. Brock. The survey team, it was expected, would contribute additional qualified scientific personnel and enrich the expedition. Stefansson did not realize that Brock — or at least some of his departmental staff — nurtured a dislike for him, and had privately voiced suspicions about his abilities. Perhaps as a result of this, an awkward "control" plan was put into effect: each scientist was given personal instructions from on high and had to report directly, through Anderson, to his specific department head. Under this arrangement, Stefansson's authority was clearly undermined, and while his original plan to explore life in the Beaufort Sea remained his to execute, the "general scientific research" he had set to be accomplished by a second Southern Party, liaising with him, was now in the hands of a team of Geological Survey scientists under the command of Anderson. And while the *Karluk* was to serve both parties, it was understood that it would operate mainly in support of the northern group, while a second ship, *Alaska*, would be made available for Anderson's party.

At the same time, as LeBourdais notes, Anderson was in the process of becoming more and more uneasy with Stefansson's personality and methods. When they were honoured at the University of Iowa together, it was Stefansson who delivered the major address, and who played the generous benefactor by donating his fee to Anderson.

It is true that there had been signs, as far back as 1910, that he was not entirely satisfied, and in spite of his friendliness after their return, it is evident that he felt he had not been fairly treated. Anderson's marriage had helped widen the rift. It soon became obvious to some of the friends to whom Stefansson had introduced the Andersons that Mrs. Anderson considered Stefansson to be a slick opportunist who was climbing to fame over her husband's shoulders. Anderson had been considered an outstanding man at the university, where he had spent all his college life, while Stefansson had been there only for a few months. Now by a single stroke Stefansson had brought about a reversal of their positions in the one stronghold of Anderson's reputation (LeBourdais 68).

On this occasion and others, the Andersons held Stefansson's easy-going manner against him, and interpreted his various acts of generosity as gestures of superiority. Perhaps because Stefansson failed to reciprocate their bad will, or even to take their enmity seriously, their resentments deepened. Even G.J.

Desbarats, from his lofty position as Deputy Minister, divined that something was amiss, and suggested that Stefansson choose a new second-in-command, a sensible idea that Stefansson blithely, and in the event, foolishly, rejected.

The expedition's members soon gathered in Victoria, British Columbia, and the first news was not good. Stefansson's chosen sea captain, Theodore Pedersen, refused to serve, fearing (wrongly) that his work for the Canadian government would deprive him of his United States citizenship. Captain Robert A. Bartlett, an experienced, crusty Newfoundlander, who had explored with Peary, was brought on, with ambiguous results, as Stefansson later decided. Nor was the group that met, mostly members of the Geological Survey, happy with Stefansson's planning: food and equipment seemed in short supply, and when more were procured, yet another ship, the *Mary Sachs*, had to be purchased to carry them.

To William Laird McKinlay, a young scientist who had come over from Scotland to join the expedition, Stefansson seemed a will-o'-the-wisp leader, hardly ever there, and making little effort to become acquainted with the men he was about to lead north. "The people of Victoria treated us like heroes," McKinlay wrote. "When we were not busy loading supplies on the ship, we were being lunched and dined by clubs and societies and private citizens who queued up to offer us hospitality. But as the days passed, I began to have misgivings about the management of the expedition. There was still no sign of Stefansson" (McKinlay 11).

Kenneth Chipman, one of the Geological Survey members, and no fan of Stefansson's, records the dissatisfaction of the research personnel following a meeting with their leader, at which they complained of various inadequacies of planning: "Murray and I were spokesmen and raised such questions as food supply, clothing, travel, equipment, where base is to be, coordination of work, etc. etc. There was nothing new! I raised these questions twice in Ottawa and his answers are the same now as then. He, however, informed us that we had no business to ask these questions, that we should have confidence in him" (Hunt 69).

It did not take very long for the misgivings of the disaffected scientists to be realized. The *Karluk* departed from Nome in July 1913, and after making a stop at Port Hope, cruised around Cape Lisburne, Alaska. Then, swept along by the ice, she headed down the coast to Point Barrow, the northern tip of the continent. For a brief spell she sailed on her own power, but was soon totally captured by the ice — and never subsequently let go. In mid-September, when the ship had been drifting eastward for a month, and was still well short of Herschel Island, fresh meat was required and Stefansson decided to take a hunting party ashore. Not long after his departure, a terrific storm carried the ship out to sea. "As the ice pack swept us further and further away from our leader, we felt not so much like soldiers sacrificing ourselves to a great cause, as lambs left to the slaughter" (McKinlay 34).

This ominous reflection was fully borne out, as William R. Hunt makes clear, largely because the *Karluk* was a ship divided. The crew was a motley one, signed by Stefansson from a distance, and there were several "green" hands among the scientists. Captain Bartlett fell out with the experienced polar explorers on board, and no coherent agreed-upon plan of action was established. On Christmas Day, 1913, the first signs of severe ice damage appeared; the ship soon broke up, and although stores were unloaded, four men were lost almost immediately in a futile attempt to reach what seemed to be the only nearby safe haven, Wrangel Island, off the Siberian coast. At that point the ship's doctor, Alistair Mackay, and three others split from the main group and decided to seek the shore instead of making for the island. They were never heard of again. After a terrifying spell of sixty sunless days on the booming, cracking ice, the remaining men attained Wrangel on March 14. Captain Bartlett soon departed for Siberia in search of help, but life on the island was marked by sickness and death, dissension, mutual suspicion, aimlessness, and lack of fresh food. One of the ship's crew, G. Breddy, died violently, apparently a suicide, though he may have been murdered. Not until September 7, when all hope seemed lost, did a rescue party arrive, on board a little trading schooner, *The King and Winge*. Bartlett, with his Inuit companion Kataktovik, had landed in Siberia, and made his way to St. Michael, Alaska, where he telegraphed the news of the ship's fate and appealed for help in the rescue. Ironically, it was Captain Pedersen, Stefansson's first choice as the expedition's chief mariner, who ferried Bartlett to Alaska, even as he had passed along the first news of the *Karluk*'s helpless drift through the polar seas. Eleven members of Stefansson's expedition had perished as a result of the *Karluk* disaster, yet their leader learned nothing of this until much later.

Stefansson expressed little empathy with the men who lived through this horrendous ordeal, although later he privately criticized the way Captain Bartlett had handled things. Bartlett, he thought, should not have complained about the *Karluk* in front of the Anderson geological party; he should never have taken the ship so far into the ice; and his supervision of the rescue was also faulty. He did not provision the advance parties sufficiently, claimed Stefansson, and he should never have cached supplies on drifting ice. He should have provided the men in the advance parties with an Inuit skin boat to make travel over open water possible. Stefansson also believed that Bartlett should have organized the hunting and food procurement on Wrangel Island with more skill, and instead of "grandstanding" by taking the rescue trip on himself, should have evacuated the whole party when he left in search of help. Not all of this critique can be attributed to defensiveness on Stefansson's part: Bartlett was later found guilty by an admiralty commission for putting the *Karluk* into ice, and for allowing the doctor's group to leave the main party, even though they had given the captain a letter absolving him of the responsibility.

While a few of Stefansson's objections may have had some validity, the absurdity of his criticism of Captain Bartlett's trek is demonstrated by William McKinlay's account of the conditions on Wrangel Island. He observes that while Bartlett had originally intended to move the whole group, "the physical condition of the majority of the party was so poor that to take them all on such a hazardous trip was inviting wholesale disaster" (McKinlay 91).

In his foreword to McKinlay's book, Magnus Magnusson places Stefansson (as did Peary in his subtly worded tribute to Stefansson later) with the last of the old school of explorers — ironically a group that Stefansson often criticized. Magnusson refers to the "terrible amateurism" of the Stefansson expedition and recalls "the heyday of the amateur, the time when men tried to climb Everest in knickerbockers and Norfolk jackets, when Britain was going wild over Scott's brave but tragic failure to be the first to reach the South Pole — another exercise in splendid amateurism." Clearly an exaggeration, but he also points out, and with a great deal of justice, that what was at the root of the *Karluk* disaster was "the absence of real comradeship and the team spirit that good leadership inspires." Stefansson's absence from the *Karluk* on her final, dreadful run may have been an accident, but it was an accident that reveals much about his leadership of the expedition.

With his departure from the *Karluk*, Stefansson began several years of Arctic adventures and exploration. While he roamed the North, the Southern Party worked steadily to gather the scientific information that was the goal of their enterprise. Their parallel endeavours were equally successful, but had quite different aims, and Stefansson's initial effort, as commander, to bring the two groups back together, and to refurbish his team with supplies and equipment by drawing on the store of the Southern Party met with strong resistance from them.

The meeting of Stefansson and Anderson at Collinson Point in spring, 1914, set the tone for the next few years. Stefansson wanted the Southern Party to survey the Mackenzie Delta and to accept the notion of expanding their geographic exploration of the northern coast whereas Anderson insisted that the Canadian government plans called for a scientific survey of Coronation Gulf; he offered to resign, rather than carry out Stefansson's plans. Stefansson refused to accept his resignation, but Anderson still opposed any diversion of supplies or men north to compensate for the loss of the *Karluk*. Hunt explains the situation as follows:

> In the conflict one can see the contrasting attitudes of the two men. Stefansson was not daunted by the absence of the *Karluk* and was determined to find the means of accomplishing his mission by redistributing resources and extending the government's credit. Compared to Stefansson's indomitable optimism, his antagonist revealed an abiding pessimism that was aggravated by hostility toward his

commander. Expedition resources would not permit exploration of the Beaufort Sea, Anderson declared. He insisted that the effort would only waste money and perhaps lives and would not produce worthwhile results. As he had not been too satisfied with his own work on the previous expedition, he may have felt a heavy responsibility for the Southern Party's success and a deep fear that Stefansson's demands would saddle him with a failure in the government's view (Hunt 92).

Stefansson's reaction was not to press the issue but to leave the base camp in order to go about re-equipping his side of the expedition himself. When he finally returned, he was confronted with what amounted to an open rebellion; Anderson, as commander of the Southern Party, refused to share men or supplies, and, as one member suggested, Stefansson "had better go home to Ottawa." Undaunted, Stefansson collected volunteers from the base camp team, announced his intention of going north, and ordered Anderson to send a relief ship, the *North Star*, to Banks Island to meet him in late summer at the very latest.

Stefansson left for his first long ice journey in March 1914. It was one of his most notable forays across the ice, and lasted some five months. He was accompanied by a few sturdy companions, including two old Arctic hands who had been with him for years, Storker Storkerson and Ole Andreason. Although the party had 3,500 pounds of provisions and four sledges, they managed the great feat of "living off the land" by taking bear and seal meat, which kept off any threat of scurvy as they travelled. Back at the Collinson Point base camp Anderson, having no news, assumed that they were lost. When George Wilkins set forth to meet Stefansson with a ship as promised, Anderson refused to allow him to take the *North Star*. Wilkins, after a trip worthy of a Jack London novel — one that included the rebellion of a crew that had to be subdued by violent means — arrived with the much less useful vessel, the *Mary Sachs*. He had no alternative, he told Stefansson; since it had been assumed that their leader was dead, Anderson was in command of the expedition.

During 1915, with Cape Kellett as his base on Banks Island, Stefansson made his second sea-ice journey, part of which took him across the top of the McClure Strait, across Prince Patrick Island, and back down Kellen Strait, which divides Melville Island and Prince Patrick. At the heart of the journey, in June, was the discovery of uncharted land, two islands which Stefansson claimed for Canada and named Brock and Borden, rather tactfully, after the chief of the Canadian Geological Survey and the prime minister.

Further planning required Stefansson's return to Herschel Island; he learned that the government planned to terminate the Southern Party's work in the summer of 1916, but he hoped to gain approval for further explorations and went on with his preparations.

During 1916, Stefansson's team ventured even farther afield, all the way to Cape Isachsen and beyond, to a new island, which he named Meighen Island after the prime minister who had succeeded Sir Robert Borden. This despite the spread of a disease which killed many of the dogs, including Stefansson's favourite, an event that, as he confessed, caused him to lose "a considerable part of the pleasure" in his work, while undermining his "confidence in the future." This rare moment of pessimism was compounded by some of the serious inadequacies he had perceived in one or two of his travelling companions during that year.

The last of the newly discovered northern lands was Lougheed Island, which he described as "an Arctic paradise, full of game and certainly capable of sustaining human life." But during the 1916–1917 season, and thereafter, Stefansson's luck seemed to run out; perhaps he was overtaxing his strength at last, with month after month of ice travel that involved constant dangers, a few close brushes with serious injury or death, and the strain of the endless effort to secure food and equipment for yet more forays into new land.

An ankle injury, the onset of painful hemmorrhoids, a near-fatal bout with typhoid fever and pneumonia, the death of two of his trusted men, Peter Bernard and Charles Thomsen, who were crossing Banks Island to bring him news from home, and the defection of one of his ship captains, Henry Gonzalez — the negative signs had begun to multiply. And complaints had surfaced when Stefansson contrived various delays, so that his ship the *Polar Bear*, supposedly sailing west from Herschel Island to return home via Nome, was icebound during the winter of 1917–1918.

In fact, the war in Europe, which Stefansson had learned of only late in his long northern sojourn, had naturally monopolized the attention of the Canadian government. And Stefansson's "discovery" of a few barren northern islands, places that to this day remain unknown to the majority of Canadians, was perhaps small compensation in Ottawa for the bureaucratic nervousness caused by the rupture between Stefansson and the Geological Survey team. The Canadian Arctic Expedition, certainly a triumph of human courage and of adaptive exploring techniques, not to mention an extremely valuable scientific venture into

Canada's great virgin North, was also a carnival of petty rivalries, personality conflicts, misunderstandings, false starts, and anti-climaxes. It lacked the "one big thing," the public "branding" that might have enabled its sponsors to gloss over some of the defects in its planning and execution. Lives were lost, ships sacrificed, equipment squandered, and rumours surfaced that did almost no one any credit. Things were such that, even before the publication of *The Friendly Arctic*, Stefansson's account of his 1913–1918 travels and discoveries, his part in the whole venture, was being undermined by newspaper attacks in both Vancouver and Ottawa. The *Ottawa Citizen*, for example, in a story published on November 25, 1916, alleged that Stefansson had disobeyed orders to return because he thought he could "flip his fingers at the Dominion Government." He expected the scientists of the Southern Group, to "live like Eskimos," the implication being that the "splendid work" of the Geological Survey team had been undermined by this eccentric and arrogant "Norwegian"— as the *Citizen* inaccurately designated him.

Despite such malicious carping, Stefansson returned south to a hero's welcome.

The Explorers Club of New York, for instance, elected him its president. One by one, too, various geographical societies formally recognized his services to science and the extension of knowledge by conferring on him the highest awards they had to offer. Medals were presented to him by the American Geographical Society of New York, the National Geographic Society of Washington, the Geographic Society of Philadelphia, the Chicago Geographic Society, the Royal Geographic Society of London, and the Geographical Societies of Paris. Later, a medal was also given to him by the Geographical Society of Berlin (Hanson 175).

Great explorers also came forth to honour Stefansson, including the two most famous figures in American polar history, Admiral Peary and General A.W. Greely. At the presentation of the Hubbard Medal of the National Geographic Society, Peary gave an eloquent testimony to Stefansson's achievement. "Stefansson has added more than 100,000 square miles to the maps of the region," Peary claimed. "He has outlined three islands that were entirely unknown before and his observations in other directions, the delineation of the continental shelf, filling in of unknown gaps in the Arctic Archipelago and his help in summing up our knowledge of those regions are in fact invaluable. Stefansson is perhaps the last of the old school, the old regime of Arctic and Antarctic explorers, the worker with the dog and the sledge, among whom he easily holds a place in the front rank.... His method of work is to take the white man's brains and intelligence and the white man's persistence and will power into the Arctic and to supplement these forces with the woodcraft, or I should say polarcraft of the Eskimo — the ability to live off the land itself …" (LeBourdais 143–44).

In Canada, however, Stefansson's reception was less effusive and more tinged with questions. Although he had been advised as long ago as 1912 by the great explorer Ernest Shackleton, to follow up his adventures and achievements with lecture tours, on the grounds that these would not only be lucrative, but would also sell his books and keep his name in the newspapers, Deputy Minister Desbarats found the idea of a proposed tour inappropriate. Stefansson's first order of business, Desbarats argued, should be the completion of an official report. The Southern Party, after all, was planning no lecture tours, but was actively engaged in preparing its scientific material for publication. It would look strange if nothing were recorded from the northern sphere of activity. Stefansson pointed out that the Geological Survey scientists had been salaried employees during their time in the North, while he had not, and that some of his lectures were being done for charitable causes. Desbarats, unrelenting, complained that Stefansson had disobeyed orders in not concluding the expedition around 1916, and Stefansson replied that he had never received those orders. Stefansson, although he agreed to cut back on his lecture tour, took the offensive by informing Desbarats of Rudolph Anderson's written threat to resign as a member, should Stefansson be elected president of the Explorer's Club for the year 1919. He requested that Desbarats require Anderson to write a narrative report for the Southern Party, and include a copy of the letter he had written to Stefansson, explaining why he had disobeyed his leader's instructions. Stefansson had in mind to present his "public narrative," soon to be published under the title *The Friendly Arctic*, as his summary of the expedition, but this of course, when it appeared, bore little resemblance to the official, bureaucratically-dressed-up document Desbarats must have had in mind. In the end, no official summary of the expedition was ever published, although Stefansson did submit a short narrative account, and various scientists, notably Diamond Jenness, published reports in their special fields.

Matters did not end there, however, for Stefansson pressed the government to clear him of any implications of incompetence or dereliction that might have arisen from Anderson's sustained vendetta. As a result, an Order-in-Council was issued, dated January 21, 1921, part of which stated:

> Mr. Stefansson's ... work and that of the members
> of his expeditions have given us much valuable

scientific information about our northern territories and seas. By developing a new method in Arctic exploration, a method of living off the country by forage, he has called strikingly to the attention of the world the fact that the North is not so barren nor its climate so hostile to comfort as had previously been regarded as true. He has thus foreshadowed an important extension of the boundaries of human habitation. In his writings and speeches he has been assiduous in calling to our attention the great national resources of the North which we had previously undervalued. He has turned men's minds towards the North Country as a possible source of food supply and a home for colonists, and his work and advice have proved the greatest incentive in promoting public and private development of our northern resources ..."

Reading such a citation, one becomes acutely aware of the ironies implicit in Stefansson's achievement and the way it was regarded in Ottawa. Unlike Peary's imperialist-style but concrete reference to the skills of the Inuit, whom he clearly considered one of "lesser breeds without the law," the Canadian government citation makes no mention of the Native people. Its strong emphasis is on Stefansson's role in opening up the North to potential development, a result that would undermine in so many ways his example of trust and respect for the Inuit way of adaptation to the environment.

Yet Stefansson himself invited this approach. As Gísli Pálsson notes, his third expedition, although it showed him exploiting to the full the survival techniques he had learned from the Inuit, was hardly focused at all on the ethnographic matters that had distinguished his previous trips to the North:

> "Stefansson's extensive published narrative of the third expedition, *The Friendly Arctic*, is more like a travel account than an ethnographic monograph. While he remains fascinated by the Inuit — particularly the Copper Inuit with whom he, again, spent some time — most of the volume deals with geographical issues, the politics of exploration, and the logistics, mental attitude, and technology necessitated by travelling on ice and in extreme cold. The Inuit, in a way, have been removed from the centre stage and the Arctic, however 'friendly', remains a natural space to be explored, conquered, and domesticated by Western 'civilization'" (Arctic Web 5–6).

The point I am making here, following Pálsson, is only apparently paradoxical, and it is, namely, that Stefansson was the worst enemy of his own future reputation as an anthropologist. The lay reader of today who is familiar with the writings of such notables as Margaret Mead, Bronislaw Malinowski, Franz Boas, Ashley Montagu, Paul Radin, and many others, can see Stefansson's best chapters on the Inuit as somewhat comparable to some of theirs in their various fields, and he is sometimes credited as being one of the early practitioners of the "participant-observation" mode of anthropology. Despite this, his work is not included in even such a standard Canadian reference book as *Eskimo of the Canadian Arctic*, edited by Victor Valentine and

Frank Valee, and published in Toronto in 1968. Scandalously, even though two books by Farley Mowat in the popular mode are listed in that book's "selective bibliography," not one of Stefansson's books or papers is cited there. This neglect, or academic hauteur, may be contrasted with the verdict of Gísli Pálsson, who writes that Stefansson "both mapped and defined the Arctic in Western discourse, paving the way for authentic accounts of Inuit society, more authentic and less ethnocentric than those previously available" (Arctic Web 4).

In retrospect, while the scientists of the Southern Party, especially Anderson, Chipman, and Diamond Jenness, can be seen to have been excessively mean-spirited and hostile to Stefansson, and to have derogated his achievements in ways that smack of the worst kind of "academic" nastiness, their discomfort was surely not altogether without foundation. The long and difficult relationship between Stefansson and Rudolph Anderson, outlined in great detail in Richard J. Diubaldo's excellent account, often reads like a scenario for a complex psychological novel, one that adds touches drawn from Jack London and Eugene O'Neill. Although Diubaldo's treatment of Anderson is restrained and balanced, even sympathetic, the reader may be forgiven for seeing Anderson's hatred and suspicion of Stefansson as often verging on the pathological. Himself ambitious, and quite gifted, but paralyzed in some respects by a deep lack of self-confidence, and unable by temperament to push himself into the limelight, Anderson resented his rival's every move and undermined him whenever he could. Stefansson, on the other hand, sometimes played the trickster and manipulator, and certainly suffered from delusions of grandeur; he was often Hamlet-like in his combination of frenetic action and passive acceptance, and his reach constantly exceeded his grasp.

As leader of the Canadian Arctic Expedition, Stefansson clearly fell short in several important respects. For one thing, the expedition was badly organized. As the accounts by William Laird MacKinlay and Jennifer Niven show, this was evident from the moment the research team gathered in Victoria. The suitability of the ships themselves, the distribution of supplies to the appropriate ships, the organizing of agreed-upon base camps, the assignment of clear and specific missions to qualified personnel, the handling of communications, the clarity of chain-of-command — all of these were areas that were sadly neglected, or mishandled, and the blame must lie with Stefansson himself. Reading all the accounts, including Stefansson's own, one gets the picture of a leader who failed to pay sufficient attention to detail, and who found no means of bridging what were really two quite different programs, the scientific study on the mainland, and the exploration of a dangerous and little known northern littoral.

Reading accounts of Stefansson's Arctic journeys between 1913 and 1918, one gets the picture of a man caught up in frenetic and impulsive activity, constantly in motion and frequently shifting his goals, of endless demonstrations of courage on the trail, but little explanation of why a particular trail was chosen, in short, of an impressive performance that somehow lacked a centre. This, no doubt, would be much less of an issue had Stefansson's part of the expedition been the only one, but given the dynamics between the northern and southern groups, one might be justified in assuming that in some sense — almost a literal one — Stefansson was "dodging the issues" that would have confronted him had he taken care to plan, to exchange information, and to supervise the work of the scientists of the Southern Party.

It is clear from everything we know about Stefansson that he was a unique individual, quite different from the skilled and talented men he worked with. In my view, Stefansson should be placed among a fascinating group of roughly contemporaneous celebrities, men of genius or near genius from various fields to whom we might apply the term "charlatan." A charlatan is literally "one who babbles" and deceives, someone who pretends to wonderful knowledge. Let us soften this somewhat, with no desire to denigrate true achievement. Let us use the word to suggest the unique character of certain creative men and their approach to their own talents, their audiences, and to the public in general. Everyone will have a personal list, perhaps including some women, but names such as Frank Lloyd Wright, Salvador Dali, Leopold Stokowski, Orson Welles, C.G. Jung, Richard E. Byrd, T.E. Lawrence, and General George Patton might spring to mind. In each of their several fields these men stood out from their often equally talented contemporaries, mainly thanks a certain self-dramatization, to the nurturing of their own public masks or personae, or because they allowed themselves to be seen as somehow "larger than life." Each one of these creative personalities has a claim to permanent attention that rivals the great or near great in their various fields, yet in each case there is something extra too, something deceptive and glittering, something contradictory and offbeat — and all the more fascinating for that — in their personalities and achievements. If we examine their public images, as well as their various works, we find what, for want of a better word, we might call "magic." Such individuals have tremendous panache and do amazing things, but are willful, eccentric, and unpredictable; they tend to divide critics and the public, and have as many detractors as they do fans.

Stefansson was clearly a virtuoso of travel. Is there an explorer an ordinary city-dweller would rather share an Arctic journey with, given that only the simplest means of survival were allowed — a sledge and some good dogs, a rifle, and a stock of ammunition? But this kind of virtuosity does not guarantee the skills of a good expedition leader. If we refer to our list of "charlatan-geniuses" above, and invoke the example of General George Patton, we see that, though he was an impulsive, flamboyant, and skilled battle commander, Patton relied on a strong organization, his own Third Army, to achieve his triumphs. This group was brilliantly organized from top to bottom, and he presided over it with keen attention and with the personal pride that a great conductor takes in the minute skills of his orchestral players. Patton and Stefansson, like some of the others on my list, had the personae of self-confident, impulsive, and eccentric creators. They could be gregarious and were socially very smooth, but were men determined to succeed, and knew how to use others for that purpose. At the same time, they were caught up to some extent in the mystique of the solitary Romantic hero or visionary. Patton, however, was part of a strong organization, which he relied on and took care to nourish. He often empathized with his men and joined with them to celebrate triumphs, and, despite some notorious gaffes (typical of the charlatan-genius), he had the skills to mould and shape various individuals in pursuit of a common goal. Stefansson, however practical and sensible, often seems oddly detached and almost indifferent to those he depended on, although he takes care to credit the achievements of certain individuals. William McKinlay's book about the *Karluk* disaster, critical as it is of Stefansson's planning and leadership, never quite brings out what he clearly feels as a deeper fault in his leader: that Stefansson never at any time seemed able to express an appropriate sorrow over his lost men. It is true that he did appear to be deeply moved by the loss, later, of his two associates

Bernard and Thomsen, but there, as in the case of the *Karluk*, he seemed perhaps over-anxious to analyze events so as to slough off his own responsibility for the tragedy.

"Stefansson's willingness to forgive was one of his finest qualities," William Hunt assures us (Hunt 149). But behind this apparent virtue is something less admirable: a kind of passive-aggression, which seeks to impose its will by seeming to yield. During the Canadian Arctic Expedition, when Stefansson was confronted by resistance and near-rebellion, he tended to retreat. When tight control and strong organization were called for, he shrugged his shoulders, or waved a hand in casual reassurance. At key moments during his time in the Arctic he was "absent, taken for dead," and however useful a strategy of disengagement may be in certain cases, it is hardly the quality of a great leader. The truth of this was borne out, not only during his last Arctic sojourn, but later, when he sought to follow up on his travels and explorations by putting some very tangible northern projects into motion. The failure of these ventures only confirms some of the conclusions we have drawn from his time as leader of the Canadian Arctic Expedition of 1913–1918.

Stefansson's Arctic companion, Fanny Pannigabluk and his son, Alex . Canadian Arctic Expedition, 1914. Glass plate.

6

Man of the World

When Stefansson left the North he settled in New York, but also spent some time in Ottawa, developing new projects and defending his reputation against the criticisms that continued to be levelled at him by the Anderson party in the Geological Survey division of the federal government.

In New York, Stefansson lived for the most part in what one might call "bachelor quarters"— at the Explorer's Club, in a hotel near the Museum of Natural History, at the Harvard Club on 44th Street, and in several notoriously untidy Greenwich Village apartments. As a devoted book collector, he amassed a sizeable library, one that tended to swamp his living space. With few domestic skills, he tended to eat out, and continued to enjoy the company of the various artists, writers, celebrities, and performers that patronised Manhattan's famous bohemian quarter.

This was an exciting time in Greenwich Village. As Gísli Pálsson notes, it was home to such political radicals as Emma Goldman, the socialist spear-carrier and women's rights advocate, and John Reed, the left-wing journalist, and author of *Ten Days That Shook the World*, whom Fanny Hurst had planned to accompany to the U.S.S.R. on his famous journey. That Stefansson was affected by the radical thought of the time seems undeniable, and he was later to suffer for it during the McCarthyism of the Cold War days.

Writers, artists, flamboyant politicians, and famous musicians were also drawn to the village, as were such public figures as Charles Lindbergh, the Wright brothers, and show business types, from actors and directors to the latest stunning Ziegfield girl. Romany Marie's restaurant was the hub for many of these celebrities, and Stefansson occasionally gave private parties in the hall above the restaurant, while attending many more at venues scattered around Manhattan.

Stefansson cut quite a figure, even among such company. As Pálsson tells us, "His name was known widely, both because he was controversial and publicly conspicuous, and also because people took note of him everywhere, a tall dignified man who stood out from the crowd" (187–88).

The Canadian government's informal discouragement of Stefansson's plan to do extensive lecture tours faded away with its realization that no authentic and official record of the great expedition would be produced by any of the parties. Although *The Friendly Arctic*, which appeared in 1921, stirred up fresh combative energy among Stefansson's enemies, its account of the 1913–1918 years was a temperate, even a rather timid one, in relation to the expedition's conflicts, and it was heavily armoured with endorsements by notables in the exploring world, not to mention a preface by Sir Robert Borden, the Canadian prime minister. Even the controversial title was credited by Stefansson to Gilbert Grosvenor, president of the National Geographic Society, who also wrote a flattering foreword for the first edition of the book.

Stefansson soon determined that his best means of earning a living was to write books and articles about the North and to travel far and wide promoting his publications and his views about the Arctic. He carried out his program in organized fashion, relying on a staff of personal researchers and editors, which cost him — as he was quick to point out — only about $5,000 a year, while his total income was between $20,000 and $30,000. Touring was another matter. As those who read biographies of musical virtuosi will recall, cross-country tours, with stays in less than opulent hotels, one-night stands, receptions, and "contact with the public," even an adoring public, can be a fairly dreadful way to pass the time. As William Hunt points out, Stefansson complained about his endless trips, and with good reason:

> Why, he wondered, could people feel pity for his suffering the so-called hardships of an ice journey and envy his life as a lecturer? Over the years he travelled to most parts of the United States and Canada, usually during the fall and spring seasons. From May 20 to June 12, 1921, for example, he delivered lectures in eighteen towns in California, four in Nevada, and one in Utah. His list of engagements reads almost like a railroad timetable: Turlock, Modesto, Lode, Stockton, Los Gatos, Richmond, Petaluma, Eureka, Willits, Fort Bragg, Ukiah,

Lakeport, Healdsburg, Sebastapol, Santa Rosa, Sacramento, Grass Valley, Reno, Lovelock, Winnemucca, Elko, and Ogden. During those twenty-three days, he slept in twenty-three different hotels, and this was just part of his season's itinerary!

To which Hunt adds:

On his lecture tours, Stefansson often stayed with friends, although when he was in Ottawa he preferred to stay at the Château Laurier — the grand old hotel that dominates the downtown — because of its proximity to government offices. He thoroughly enjoyed being entertained, so he did not share a fellow lecturer's aversion to social life, although he could appreciate his point of view. Elbert Hubbard, a well-known writer, at one point teamed up with Stefansson on the same program. On Hubbard's business card, according to Stefansson, were prominently displayed his terms for lecturing: $100 IF I STAY AT A HOTEL, $250 IF I AM ENTERTAINED (Hunt 158).

One reason that Stefansson enjoyed being "back in civilization" was that it brought him into the company of many charming women. Having parted from his first real sweetheart, Orpha Cecil Smith, around 1920, he was open to both casual liaisons and to more serious relationships. His lectures created an easy path of contact, and he had numerous devoted admirers, many of whom wrote him constantly, collected clippings of his travels, and threatened to include him in their creative efforts. Peggy Fletcher, a would-be novelist, gushed to him, "I would like to make a living man step out of the pages — a man as arresting and full of mystery as you are" (Hunt 159).

It is not clear how many of these women knew about Stefansson's past, in particular that he had cohabited with Pannigabluk, a seamstress, one of his main Inuit companions during his Arctic years. She had borne him a son in 1910, and groping for ammunition to use against Stefansson, some members of the Southern Party had condemned him for refusing to acknowledge paternity and for neglecting his Native wife and son. Gísli Pálsson, who has investigated this matter fully, concludes that Stefansson probably invited his son, Alex, to leave the Arctic and come south, but may have changed his mind later. Pálsson notes that, given the bigotry and racism that abounded in the United States (and Canada) in the 1920s, Stefansson would have been loath to expose Pannigabluk to such an ordeal. He also suggests that Pannigabluk might well have been unwilling to part with her only son, "who was her assurance of a livelihood in old age" (172). Interestingly, Pálsson points out that if Stefansson did in fact finally refuse to take Alex under his wing, the experience may have been coded in *Kak, the Copper Eskimo*, a young adult novel co-authored by Stefansson with the novelist Violet Irwin and published in 1928. In the story an Inuit boy is befriended by a white visitor; he has fond dreams of accompanying his older friend south, but in the end is left behind. As the authors describe the parting, "Talking about impossible dreams as if they were about to happen makes them seem jolly real. Kak managed to choke back his sorrow and, freshly convinced that life was a great adventure, ran after the party who were already trekking north" (*Kak* 231–32).

Pannigabluk died in 1940; interestingly, Stefansson was married for the first time in 1941. Alex Stefansson pointed out later that his father did not marry until his mother was dead, proof perhaps that he felt some of kind responsibility toward her. June Helm, an Iowa professor who had accompanied her husband to the Arctic, provided Gísli Pálsson with considerable information about Alex Stefansson. She explained that his connection with the explorer was known locally, and that he himself seemed proud of his relationship with the famous white man, although mixed-race children were often subject to harassment by their Inuit brethren. It also seems fairly well established that Stefansson, contrary to the gossip of his fellow explorers, did provide material support for his Inuit family. The fact that Stefansson did not disclose that he had a son, Helm found not surprising, given the prejudices of the era. For a well-known personality to admit to having "abandoned" a child in the North would have been to invite gossip, scandal, and moral censure. There was no easy option, and on balance, Stefansson's decision to take no action and keep silent is certainly understandable.

Stefansson managed to avoid any serious commitment with most of the women who pursued him, but during the 1920s and 1930s he found a great love in the popular novelist Fanny Hurst. Born in the American Midwest into a Jewish manufacturing family, she moved to New York, where she did some post-graduate study, and published her first book in 1914. A string of novels and short stories followed, and by 1940 she was the highest paid writer in the United States. She was also a striking personality, known "for her daring behaviour, striking clothing and unusual jewellery.... She is said to have boasted that she had slept in all the rooms of the White House, with the exception of the bedroom of the President and First Lady.... Her ideas of the "new" woman, woman's struggle in a male-dominated society, and relations between the sexes sound surprisingly relevant, even to our ears today. A recent work on New York life during the 1920s suggests that her works are a rich source of information about contemporary views on women as sexual beings" (Pálsson 190–91).

À la mode as it was during her day, her work has since become outmoded. In retrospect her fiction has been judged to be of the soap opera variety, and she has been called "a sob sister" writer who relied on exaggerated metaphors and a generally overblown style to get her effects. Nonetheless, significant films have been made from her stories and one of them, *Imitation of Life*, about a woman's rise in the business world, was twice successfully made, with important Hollywood stars in both versions, and with the well-known Douglas Sirk directing the second effort.

Hurst and Stefansson did not, however, meet in Greenwich Village, but in Ravello on the Amalfi coast, in Italy. He was forty-three, ten years older than she was. Theirs became one of those intense, consuming passions — especially on her part — that occur on the literary landscape from time to time. It seems to have been strong on many levels, physically compelling (she found him to be a passionate lover), psychologically engrossing, and intellectually stimulating.

Hurst was married to a professional musician, but, anticipating the Woody Allan-Mia Farrow relationship, at least in one respect, they lived in separate New York houses or apartments, a practice which a *New York Times* editorial denounced as an unnecessary extravagance during a housing shortage! The Hurst-Stefansson relationship lasted some seventeen years, and in 1932 she wrote him a strikingly extravagant and almost Byzantine birthday greeting: "Dearest, Fifty-three years ago today the loveliest and most important event of what was to be my life preceded me by ten years. You are still my most important event. You always will be. Be happy, and bless you, bless you, bless you" (Pálsson 192–97).

Perhaps the most amazing thing about this relationship, given its intensity and endurance, is that the lovers managed to keep it secret from almost everyone. If it had not been for the accidental discovery of some boxes of the explorer's letters at a flea market in New England in 1987, long after Stefansson's death, it probably would have remained unknown forever.

7

Reindeer, Wrangel Island Again, and the Last of Canada

The Friendly Arctic is indeed a friendly, upbeat book, one in which Stefansson takes pains to be fair to those he knew disliked and suspected him. These included Captain Robert Bartlett, who, as Jennifer Niven notes, could barely pronounce the explorer's name without salting it with some good old-fashioned Newfoundland expletives. It also comprises most of the "Geological Survey gang," who worked very hard after the book's publication to discredit it, as well as some members of the ill-fated *Karluk* adventure.

With shrewd foresight, Stefansson had obtained some advance blessings for his vivid but rather eccentric account of the Canadian Arctic Expedition. As I have indicated, testimonials from Gilbert Grosvenor of the National Geographic Society, Admiral Peary, and General Greely, as well as a full-blown introduction from Prime Minister Sir Robert Borden, preceded

Stefansson and others aboard the Karluk.

the actual text of the first edition. This obviously made it very difficult for Stefansson's detractors to gain much momentum in their efforts to discredit the book, and it is almost certain that Stefansson calculated his narrative rather carefully, telling a few home truths along the way, but not adopting a tone of outright hostility at any point. In fact, he was a master at conveying his "understanding" of certain actions taken by others, never characterizing them as "wrong," but at the same time bringing forth his own cogent reasons why his ideas, or the course of action he would have chosen, might have been better. In *The Friendly Arctic*, for example, Stefansson refrains from calling Rudolph Anderson's resistance at their Collinson Point confrontation "mutiny," but he was much more blunt and critical in letters to Ottawa and elsewhere. In the pages of his book, on the other hand, "Stefansson's references to Anderson's performance were carefully phrased and unlikely to arouse curiosity in the minds of readers lacking inside knowledge" (Hunt 177).

Among some readers with inside knowledge, however, it was a different story. Several expedition members from the government's Geological Survey team complained about Stefansson to their superiors for many years "before they were ordered to refrain from a public discussion that might prove embarrassing to the government" (Hunt 179). Although the long Conservative hegemony of Sir Robert Borden and Arthur Meighen was about to end and Mackenzie King's Liberals were ready take power, the policy and the government message remained the standard one: "avoid embarrassing us by airing your nasty disputes in public!" This did not prevent the disaffected scientists, however, from spreading gossip on social occasions, from passing along loaded information to newspaper editorial writers, or from working to block honours coming to Stefansson in Canada and elsewhere.

When, about this time, Stefansson was nominated, by Sir Robert Borden no less, for membership in Ottawa's prestigious Rideau Club, it seemed a foregone conclusion that he would be accepted. When he was blackballed, and, in typical fashion, apologized to the former prime minister for causing him unnecessary annoyance, Sir Robert wrote back: "Pray do not feel any regret on my account. My regret is that the Rideau Club has made itself so supremely ridiculous. In my opinion it was honoured in having you seek admission as a member" (LeBourdais 148). In 1960, two years before Stefansson's death, the Rideau Club, extending too little too late, granted him an honorary membership, although the privilege had previously been conferred only on a few governors general of Canada.

Meanwhile, an editorial in the *Montreal Standard* blasted Stefansson as a showboater and opportunist who was marketing his expedition story to the magazines, instead of setting down his research in sober and scholarly publications. "An imported adventurer," he was subverting the work of Canadian scientists and was acting "more like an exploiter than an explorer." The paper called for a government inquiry on the grounds that the great expedition had cost "ten times more than the estimate" (Hunt 179).

This piece bears all the trademarks of a canard dictated almost directly by Rudolph Anderson, for the references to magazine sales had been a sticky issue between Stefansson and Anderson from the first, and it was typical of Anderson and his team to denounce Stefansson as a showboater and a charlatan. Although the latter estimate was a judgment not altogether without foundation, to pass over Stefansson's good qualities in the interest of a cheap put-down was to distort his achievement in egregious fashion.

No such mistake was made by the *New York Times*, which, after reporting the Canadian controversy in its news pages, took up the case in an editorial. As the editorial writer pointed out, disputes among those who return from dangerous expeditions are nothing new; they are to be expected when survival has been at stake, and controversial decisions made. The *Times* suggested that it was clear to all who knew him, and should be obvious to his readers, that Stefansson was honest, courageous, and competent. Perhaps Stefansson, cast in the Viking mould, would arouse antagonism in those with a distaste for discipline. If so, as the writer concluded, "the case as it stands is against *them*" (Hunt 183–85).

More subtle than the newspaper attacks was a long letter in the magazine *Science*, published in 1922 by Diamond Jenness. He was responding to a favourable review of *The Friendly Arctic* which had appeared in a previous issue, written by Professor Raymond Pearl, who claimed that the Southern Party, resisting Stefansson's orders, had believed him to be "not merely silly but probably also insane" (Hunt 180).

In replying, Jenness conveyed the impression that the government had vested leadership of the Southern Party in Anderson. This was not true, and was corrected, at Stefansson's request, by the author of the orders himself, G.J. Desbarats. Not only did Desbarats reaffirm that the government had given Stefansson overall command of the expedition, he pointed out that Jenness had blurred the intent of the instructions by his manipulation of the text. Desbarats was not the only former or active high government official disturbed by the backbiting initiated and sustained by the scientists. R.W. Brock, who had been head of the Geological Survey during most of the expedition time, sided with Stefansson, at least to the extent of assuming that the explorer wished to make peace with the scientists from his former expedition team.

Stefansson had taken care to try to preserve the considerable good will that existed toward him among several high Canadian government officials, and seemed to wish to continue his activities in the North, using Canada as a springboard. On May 6, 1919, he addressed a joint meeting of the Senate and House of Commons, on the subject of the Canadian and North American food supply. On that occasion he presented a grand picture of what benefits might result from the government's fostering of a large food and wool industry in the North based on the creation of vast herds of reindeer and musk-oxen (because of the derogatory connotation of *musk* in relation to food, he preferred to call the musk-oxen *ovibos*). "If we do this on a large scale," he argued "we shall through these two animals within the next twenty-five years convert northern Canada from a land of practically no value into the great permanent wool, milk, and meat producing country of the western hemisphere" (LeBourdais 152–53).

A few weeks later, Arthur Meighen, at the time head of the Department of the Interior, appointed a Royal Commission to investigate the whole question, one that convened only in 1920, by which time Stefansson was already pushing forward with his own commercial plans in the North. He had approached the Hudson's Bay Company with his reindeer breeding plan, and applied for an exclusive grazing lease covering 182,000 square kilometres of territory on Baffin Island. By June, 1920, the lease had been approved, and was given Meighen's blessing. The catch was that at least one thousand reindeer had to be on the land by November, 1924, and six thousand by November, 1932, with an eventual total of ten thousand head. Native caribou were to be incorporated gradually into the herd. Storker Storkerson, one of

Stefansson's former Arctic travelling companions, was hired to do the initial survey of the land and to purchase the reindeer from Norway.

Typically, Stefansson was occupied elsewhere when the plan began to fall through. Storkerson, whose initial report on the foraging possible on the island was later criticized as misleading, resigned when it turned out that he would not have full control of the operation. Less than six hundred reindeer were delivered and the herd dissipated over the next few seasons, some having died, others absorbed by the caribou herds. The lease was soon cancelled and the company dissolved. Expert opinion has since concluded that the choice of Baffin Island was probably a mistake, but Stefansson, who had been unhappy with some of the Hudson's Bay Company's arrangements for the project, argued then and later that the choice of manager and the selection of the Lapp herders played a major part in the fiasco (Diubaldo 157–60).

⁂

Stefansson's next major venture in the Arctic was his attempt to colonize Wrangel Island, the bleak refuge, 177 kilometres north of Siberia, from which a handful of the *Karluk* survivors had been rescued some years before. His choice of Wrangel to expand his activities seems strange. One might imagine that after being associated, even indirectly, with a catastrophe of the dimensions of the *Karluk's*, Stefansson would have thought twice about trying to connect with the place that had such associations of suffering and death for many of his former shipmates and fellow explorers. Or, from another perspective, that some inbred Norse sense of caution might have alerted him to the fact that he was, in a sense, tempting fate. It is even possible, according to Richard Diubaldo, that Stefansson slightly doctored the written account of the island by Jack Hadley, a *Karluk* survivor, to make Wrangel appear to be a "demi-paradise" (Diubaldo 162–63).

Clearly a kind of hubris was operating; the North was Stefansson's territory, and his hyperactive nature required some new venture there. "To Stefansson's way of thinking, Wrangel Island offered a stepping stone to the undiscovered riches that the polar basin had to offer" (Diubaldo 164). And, short of actually visiting the place, he would do everything in his power to get his plans for the island accepted by other parties: the Canadian and United States governments, the Hudson's Bay Company — it really didn't matter. Such determination was part of Stefansson's romantic drive, the drive to achieve the impossible, but it was not purely that: he also needed a good investment, and hoped to make money from the venture.

Stefansson had enough influence with Sir Robert Borden, the former prime minister and with Arthur Meighen, his successor, to get his project going. And the situation was promising. In the fall of 1920, Stefansson convinced Meighen that Canada was confronted with the necessity of asserting itself in the North, and that the two sensitive touch-points were Ellesmere and Wrangel Islands. But upon advice from some of his senior ministers, who doubted the commercial value of Wrangel and had misgivings about the international implications of asserting a claim there, Meighen backed out. He then considered placating Stefansson by offering him the chance to lead an expedition to Ellesmere, but Sir Ernest Shackleton, the renowned Antarctic explorer, had already found backing for such a venture, including $100,000 from Sir John Eaton, the department store magnate, and the cabinet could not decide to go with either explorer. Stefansson felt that Shackleton, with whom he had previously discussed possible northern expeditions, had double-crossed him by indicating to the government that Stefansson had no further interest in going north. "Many prominent Canadians," he wrote, with splendid irony, in his autobiography, "felt that a second-rater like me should not be supported by the government when it was possible to get an authentically great man in my place." And those suspicious of him as "a faker and charlatan," he thought, had worked against him in the cabinet. Since it was not long after that time that Stefansson was unaccountably blackballed for membership in the Rideau Club, this was no mere paranoia.

Having failed to draw the government into his plans for Wrangel Island, Stefansson decided to go ahead on his own. He created a company for the purpose, with the conviction that once he had established his team on Wrangel, and officially or unofficially claimed it for Canada, the government's hand would be forced. Stefansson himself did not lead the expedition; it was instead commanded — at least on paper — by Allan R. Crawford, a twenty-year-old science student at the University of Toronto, whose recruitment for this position was based on the fact that he was a Canadian. The other expedition members, Arctic veterans Fred Maurer and Errol (Lorne) Knight, and an adventurer from Texas, Milton Galle, were Americans. The team sailed from Nome, Alaska, on September 9, 1921, and as soon as they arrived at Wrangel, ran up the Union Jack and claimed the territory for Britain. Public and private repercussions followed. The Canadian, British, and American press all reacted negatively: the Canadians asked, why Wrangel, of all places? The Americans suggested that the territory, strategically, was far more vital to United States interests. The British worried that this "wildcat" claim might damage Anglo-American relations.

A series of complex, near Byzantine, diplomatic negotiations ensued, all carefully detailed in Richard Diubaldo's exhaustive account. The upshot of all of this was that the British, fearing a negative United States reaction, backed away. The Americans, although they considered the possibility of establishing an air base on Wrangel, finally also disengaged. Both countries preferred to let sleeping dogs lie. Meanwhile the Soviets, fairly recently established in old Russia, considered the possibility of staking out their own claim to the island, which was, after all, reasonably close to the Siberian coast. Although it had been visited by British and American whalers, and by Russian sledge parties in the nineteenth century, the island had been named after a Russian governor of Alaska, Baron Wrangel.

The four Stefansson-inspired adventurers, and their single Inuit companion, Ada Blackjack, a seamstress, reached Wrangel on September 15, 1921. The next seventeen or eighteen months passed reasonably well, although the re-provisioning ship did not reach them during the first year. The bleak Arctic, however, slowly took its toll on the whole party. Their hunting and fishing were not varied and successful enough and they had to move camp to obtain sufficient fuel to warm them. When, in January 1923, as part of the pre-planned communication strategy, Knight and Crawford started a trek to Siberia to reach a telegraph post, they had to turn back because of illness. Galle, Maurer, and Crawford then made the attempt. Tragically, they disappeared completely, apparently having drowned when they fell through thin ice en route. Left on the island with Ada Blackjack, Knight survived for a while. He kept a journal and was fed by the Inuit woman, who hunted, fished, and did all of the work in camp.

Stefansson, meanwhile, was struggling to organize a rescue operation. He finally succeeded in obtaining private funds, mostly from Britain, and an old trail-mate from the Canadian Arctic Expedition, Harold Noice, ended up leading the relief group. He was not Stefansson's first choice, but it was not easy to get a ship from Alaska to enter Siberian waters, since the Soviet government had requested that any rescue ship stop first in Siberia, not a port-of-call most of the available mariners were eager to make. In late August 1923, Noice returned to Nome with the bad news, which was enormously amplified by his subsequent articles, many of which were of a sensationalist nature. He wrote of men starved and drowned as they desperately trekked across the dangerous ice to Siberia in search of food, and of Ada Blackjack nuzzling up to the dying Lorne Knight and refusing to get food for him unless he married her. And some garbled versions of his interviews produced even more sensational versions. He also appropriated Knight's diary, and apparently removed some of the entries, confusing still more the picture of events on the fateful island.

Noice eventually met with Stefansson and recanted some of his more lurid revelations about the tragedy, but it was too late. Article after article had appeared, and nothing in them was calculated to enhance Stefansson's reputation, or to do anything but provide ammunition to those who desired to ridicule his favourite notion of "the friendly Arctic." Canadian newspapers condemned the explorer, and the story flashed around the world. In fact, Stefansson was hardly the only one responsible for the tragedy; some of the men in the party were experienced explorers who might have been expected to do better; and they disregarded some of Stefansson's good advice. And, although Allan Crawford's shocked and grieving parents blamed Stefansson for the disaster, Lorne Knight's father supported him throughout.

Stefansson soon sold out his interest in Wrangel Island to Carl Lomen, "the reindeer king," an Alaskan millionaire, but in August 1924, a Soviet warship, the *Red October*, arrived at Wrangel Island and removed the twelve Inuit and the American trapper who had been deposited there by Noice. Lomen received compensation from the United States government and did fairly well by the transaction. Stefansson got nothing from Britain or Canada.

The Wrangel Island disaster dealt a near-mortal blow to Stefansson's reputation in Canada. Although Prime Minister Mackenzie King had not acted decisively to deal with the situation, he had not enjoyed being brought close to an international crisis by Stefansson's obsession with claiming the island for Britain (and indirectly, for Canada). And naturally the government had distanced itself as far as possible from the grim end of the Wrangel settlers, and the subsequent furor, which had given Stefansson's enemies, in particular Rudolph Anderson and his scandal-mongering wife, plenty of ammunition in their lifelong feud with Stefansson.

As Richard Diubaldo sums up the situation: "There was no deliberate government policy to ostracize Stefansson, but, as he became more insistent in his demands and activities in Ottawa between 1918 and 1923, it became clear that government co-operation with him must be predicated on Stefansson's toning down his manner and softening his posture. It was not in Stefansson's nature to retreat, or to subordinate himself to others. What must be done was obvious to him: forsake his Canadian career" (Diubaldo 209).

Although born in Manitoba, Stefansson, as we have seen, had moved to the United States at an early age and had grown up and been educated there. In his mature life, he was far more connected with the eastern United States establishment, in particular the intellectuals, artists, explorers, scientists, and journalists, than he ever was with the rather subdued and semi-invisible Canadian elite. He had lived for a long time in New York, and at the height of his Canadian activities was still travelling across the United States, growing famous as a lecturer, and earning a considerable income from that activity and from publication in well-paying American magazines, such as *Harper's*, *National Geographic*, *Collier's*, *The American Mercury*, *Redbook*, and *The Saturday Review of Literature*. When things did not work out in Canada, he did the sensible, indeed, the obvious thing: he turned back to his American roots.

Kamanna skinning an Ugyuk or bearded seal near Little Whale Bluff, Franklin Bay, Northwest Territories, June 21, 1912. Stefansson expounded the benefits of an all-meat diet when he returned to New York.

8

Marriage and a Matter of Diet

In the late 1930s, someone at Romany Marie's inimitable Greenwich Village restaurant, perhaps even the colourful proprietress herself, introduced Stefansson to a Jewish woman from Brooklyn named Evelyn Schwartz. Evelyn was dark-haired and attractive, a very "modern" woman, in some ways a similitude of Fanny Hurst, but younger and still trying to find a career. At the time, she worked with her husband, Bill Baird, who ran a marionette show; she also sang occasionally in clubs and restaurants and was studying fine arts. Evelyn and Stefansson were attracted to each other from the first, but their relationship seems to have progressed in a rather leisured and sedate fashion. When he ran across her on a New York street in 1939, he learned she was looking for work, and recommended her for a job at the Icelandic pavilion in the famous World's Fair. By this time, Evelyn and her husband had separated and she and Stefansson began to meet regularly and not very long after got married. He was sixty-two and she only twenty-seven.

Their union surprised Stefansson's friends, who had assumed that he was a confirmed bachelor, but, by all accounts, it was a most successful partnership on both the personal and professional levels. Evelyn became a skilled writer, editor, and librarian, and had many working years following Stefansson's death in 1962. She trained in and practised psychotherapy, learned French, married for the third time — a wealthy academic — and capped her successes by becoming a serious art collector and philanthropist. Whereas in Stefansson's relationship with Fanny Hurst, Fanny seems to have been the doting partner, with Evelyn the situation was reversed. "I love you beyond anything I have experienced, up to the limit I can conceive," the usually careful Stefansson wrote her at one point (Hunt 255).

One of the incidental but fascinating subjects that Stefansson explored on a personal level during his years of residence in New York City was the human diet. As I noted earlier, he had been interested in this subject from his youth and was one of the few northern explorers (Knud Rasmussen, who was part Inuit, was another), who lived almost wholly within the framework of Native languages, customs, and provision while on the trail. Almost from the first, Stefansson adopted the Native diet of fish and meat, without salt, and often uncooked, casting away the notion of hauling canned goods, or subsisting on the traditional explorer's diet of fat-enriched pemmican.

Having experienced no ill effects from this regimen, Stefansson took issue with the medical dogma that the only reasonable diet for a "civilized" person was a varied one, with a strong component of fruits and vegetables, as many as possible eaten raw. As for red meat, early-twentieth-century science had already tagged it as a suspect food — one that if indulged in too freely might cause high blood pressure or liver damage. The experts of the day therefore advised moderation in its consumption. The idea that a diet composed exclusively of meat and fish could be a healthy one seemed the notion of a crackpot, and that the Inuit were known to subsist well enough on such a diet was attributed to special physical and environmental factors.

In the years after his return from the Arctic, Stefansson had undergone a series of medical examinations. One doctor summed these up as follows: "In 1925 a number of my New York medical colleagues and I made a thorough survey of Stefansson.... We found no evidence of the ill effects which a good many at that time expected in a man of forty-six who had been anything from a moderate to a heavy meat eater all his life, and who lived on literally nothing but meat and water for an aggregate of several years" (Hanson 225).

In 1928, as a follow-up to this, Stefansson and his former trail-mate Karsten Andersen, who had been living in Florida and had in the months just prior been periodically unwell, agreed to serve as guinea-pigs for a new experiment in diet, carried out by the Russell Sage Institute of Pathology in New York, and conducted at Bellevue Hospital. As Stefansson himself explained in his 1935 article "Adventures in Diet," published in *Harper's* Magazine:

> The aim of the project was not, as the press claimed at the time, to "prove" something or other. We were not trying to prove or disprove anything; we merely wanted to get at the facts. Every aspect of the results would be studied,

but special attention would be paid to certain common views, such as that scurvy will result from the absence of vegetable elements, that other deficiency diseases may be produced, that the effect will be bad on the circulatory system and on the kidneys, that certain harmful micro-organisms will flourish in the intestinal tract, and that there will be insufficient calcium.

The researchers working with Stefansson and Andersen took some care to exclude commercial exploitation of the tests. The American Institute of Meat Packers, seemingly confident of their product, had asked permission to reprint the results, but as Stefansson explains: "We refused permission to reprint, but suggested that they might get something much better worth publishing, and with the right to publish it, if they gave a fund to a research institution for a series of experiments.... We gave the meat packers warning that, if anything, the institution chosen would lean backward to make sure that nothing in the results could even be suspected of having been influenced by the source of the money." After much negotiating, the Meat Packers Institute agreed to finance the Bellevue experiment, and later bought and distributed Stefansson's book on the subject. The trials lasted a year. During the first six weeks the subjects were under constant supervision, and during the first six months they slept at Bellevue. For the final six months Stefansson and Andersen were allowed more freedom, but agreed to keep to the conditions of the diet. The results were summed up by Stefansson as follows:

The broad results of the experiment were, so far as Andersen and I could tell, and so far as the supervising physicians could tell, that we were in at least as good average health during the year as we had been during the three mixed-diet weeks at the start. We thought our health had been a little better than average. We enjoyed and prospered as well on the meat in midsummer as in midwinter, and felt no more discomfort from the heat than our fellow New Yorkers did. ("Adventures in Diet")

It would be risky, however, to draw from this too many conclusions about the relative worth of carbohydrate, meat, and vegetable components in twenty-first-century eating plans, or to suggest that the Stefansson experiment lends credibility to recent low-carb fad regimens, such as the Dr. Atkins or Suzanne Somers diets. For one thing, while existing on meat and fish in the North, Stefansson and Andersen were getting exercise that would leave most of the fanatic joggers and cyclists of today in the dust. And there is no doubt that the fish and meat they ate, both in the North, and in the experiment, were a great deal better than most of the fast foods, snack foods, and pre-prepared microwave foods that so many North Americans still consume in huge quantities, and probably better than most of the meat and fish available from all but the specialized organic suppliers today.

Stefansson did insist, however, that his favourite meat diet would be useful in losing weight. He wrote:

I was about ten pounds overweight at the beginning of the meat diet and lost all of it. This reminds me to say that Eskimos, when still on their native meats, are never corpulent — at least I have seen none. They may be well fleshed. Some, especially women, are notably heavier in middle age than when young. But they are not corpulent in our sense. When you see Eskimos in their Native garments you do get the impression of fat round faces on fat round bodies, but the roundness of face is a racial peculiarity and the rest of the effect is produced by loose and puffy garments. See them stripped and you do not find the abdominal protuberances and folds which are numerous at Coney Island beaches and so persuasive in arguments against nudism. There is no racial immunity among Eskimos to corpulence. You prove that by how quickly they get fat and how fat they grow on European diets. ("Adventures in Diet")

Stefansson never ceased believing in his anti-carbohydrate crusade. This could lead to some social awkwardness. Hostesses could be irritated, we are told, "if Stefansson should decide, as he sometimes did, that a buffet dinner offered too many carbohydrates. He would ask for a quarter of a pound of butter and a spoon, and would eat it unperturbed; he would have been astonished to learn that he had offended anyone" (Hunt 262). In restaurants, he would insist on the bacon being "only slightly warmed," with no fat cut off, and would instruct the waiters to "bring extra pats of butter but no potatoes or vegetables" (Hunt 262).

Stefansson's views on diet, deriving from his Icelandic background, some of his university studies, and above all from his Arctic experience — although they have some relevance to our present obsession with nutrition and health — remain an eccentric and distinctly minor aspect of his intellectual bequest to our century.

Stefansson dragging a seal. This photo was used as the cover image for his book The Friendly Arctic.

9

Trials of a Different Order

S tefansson's world was changing. In the years leading up to the Second World War, the private sector and the United States military began to draw upon his knowledge of the Arctic. As early as 1932, he had been a consultant to Pan American Airways, which, for several years had run polar flights, and in 1935–36 he was commissioned by the Army Air Corps to produce a guide book and Arctic manual, which went through several revisions and printings right through the war years. In 1941, at the request of the United States Navy, he compiled several volumes of sailing directions for northern sea routes. With the country fully at war, he served as a frequent consultant to various army units and divisions, including the High Command of the Alaskan theatre of operations. He flew everywhere, to the West Coast and Colorado, to Labrador and Baffin Island, an expert on everything from cold weather clothing to diet, transportation, and survival techniques.

By far the most ambitious war project undertaken by Stefansson, however, was the planned *Encyclopedia Arctica*, initiated by the United States Office of Naval Research, and intended as a twenty-volume, five-million-word compendium of knowledge covering all things Arctic. In 1946, Stefansson was contracted to edit this impressive compilation, while his long-time secretary-researcher Olive Wilcox was to manage the project. With generous funding from the Navy ($200,000 for the first two years), the encyclopedia would include scholarship and expertise from contributors around the world, and range over many disciplines.

Stefansson was surely the perfect choice as editor of the encyclopedia. Not only was his on-the-ground familiarity with the region nearly unparalleled, he was vastly well-read on the subject, and was also a good linguist and writer, and a man with innumerable personal connections in nearly all areas of Arctic knowledge. To top it off, as a collector with a world-class library of books on the Arctic, Stefansson had many resources and much information at his fingertips. His library, D.M. LeBourdais informs us, was the world's greatest private collection of books on the Polar Regions and associated subjects (and second in importance only to that of the Institute of the Arctic in Leningrad). Stefansson had begun his Arctic book-buying in 1929, and as Olive Wilcox wrote: "When Stef sets out to collect a library he collects a library!" (LeBourdais 184).

The books overflowed two New York apartments, and spread to his farm in Bethel, Vermont. "Horse stalls make very good book bays," Stefansson told LeBourdais, on one of his friend's visits. Stefansson opened his New York collection to interested students, and by the time all the sections were brought together and donated to the Baker Library at Dartmouth University, the total comprised some 25,000 volumes, 20,000 assorted pamphlets, and many rare manuscripts.

Despite this treasure-trove of resources, and despite Stefansson's experience and connections, however, the Arctic encyclopedia never quite got off the ground. Or rather, it had just taken off, and seemed most promising, when, in 1948, the U.S. Navy — abruptly, and without adequate explanation — pulled the plug on the project. By then, two volumes had already been completed, but not published, and others were near completion. Various reasons have been suggested for this about-face. One possibility is budgetary limitations: the Navy had just funded, launched, and prematurely cancelled its Operation Highjump, the largest Antarctic expedition ever, led by polar explorer and navy veteran Richard E. Byrd. It included a flagship, an aircraft carrier, plus thirteen navy support ships, and nearly thirty aircraft of various kinds. While the two projects were incommensurable in terms of financial commitment, whatever factor called a halt to the Byrd expedition may also have short-circuited the Stefansson project — odd logic often rules when governments decide to trim budgets.

Perhaps a more likely explanation of the cancellation of the encyclopedia project, however, is the one advanced by several of Stefansson's biographers. They suggest that the explorer's pro-Soviet stance, maintained well into the Cold War period, got him into trouble with the U.S. Navy. The Navy carefully screened all those who worked on the encyclopedia, using monitors or "checkers" to ascertain the "loyalty" and reliability of the participants. And when they began to weed out Soviet specialists and American experts trained in the U.S.S.R., Stefansson took the offensive. As politically bludgeoning as ever, he pointed out that he himself had praised the Soviet Union's activities in the North. He made jokes about the Navy's suspicions and, in another act of amazing, but no doubt foolish, bravado, put the case to General George C. Marshall, the U.S. Army Chief of Staff. Surprisingly, Marshall supported him, which surely did nothing to endear Stefansson to the Navy (Hunt 257–58).

Although never a member of the Communist party, Stefansson's sympathies were clearly with the left, and in particular he was an admirer of the Soviet Union. He became neither an outright "fellow traveller," such as the great American singer/actor Paul Robeson, nor a sympathetic and curious visitor, such as the writer H.G. Wells, but his books had been translated and were well received in the U.S.S.R., and he acknowledged this and declared himself proud of it.

Stefansson's most interesting commitment to Soviet social experiments was his involvement in the so-called Ambijan Committee. This was a group founded in the United States in 1934 to support the Soviet plan to create a "socialist autonomous region" for Jews in Birobidzhan, in eastern Siberia, on the Chinese-Manchurian border. This area was, if one accepted the propaganda, to be a refuge for Jews persecuted under the old European regimes, and threatened by the rise of the Nazis. It was designed as a kind of new Zion, one that preceded the founding of Israel, and it did in fact, in due course, allow some refugees from the Nazi terror to escape the Holocaust. A more cynical view suggests that the gathering of Jews in this obscure area of Siberia by Stalin was designed to make it easier for him to eliminate them, if he so chose, but this notion was not prominent during the "honeymoon" era of relations between the U.S.S.R. and its American sympathisers (Pálsson 212–15).

Stefansson was a prominent member of the American Ambijan Committee, which included other notables, such as Albert Einstein, and after the Nazi-Soviet Pact broke down and Hitler invaded Russia, he lent considerable energy to the cause, making speeches, writing letters of support and attending meetings. The American groups continued to work in this cause until 1949, when the Cold War intervened. In retrospect, it turns out that the Soviet experiment was heavily flawed. Many Jews were purged, working conditions were difficult, ancient rites modified or abrogated, yet the Jewish component in this region of Siberia has endured to the present day. Taking climate alone into consideration, never mind the issue of "homeland," it is clear that such a resettlement could hardly appeal to world Jewry in comparison with the lure of Palestine, yet Russian and American commitments to this project were in many cases quite sincere, and Stefansson's — thanks to his complex involvement with two remarkable Jewish women, and to his appreciation of the Soviet response to his books — was certainly so.

As we can see from such examples, although Stefansson's social interests were diverse, and his perspectives rather unique, in some ways he emerges as a typical intellectual of the 1930s. He sought to support "progressive" causes wherever they touched his sympathies or his expertise, and hardly suspected that his free-floating enthusiasms might soon be seen in a much more sinister light by the witch hunters of the Cold War era. Like many of his generation, he saw the Soviet Union as a country devoted to science, progress, and social equality, and in his case this perception was reinforced by the fact that, unlike Canada and even the United States, the Soviets took Arctic development and exploration very seriously.

From the first, he had made no secret of his admiration for the relatively peaceful and communal social situation he found in Inuit society — this alone, during the witch-hunting 1950s, might be grounds for suspicion. Stefansson was a generous liberal soul of the old school, and was shocked, as were many of his ideological stamp, when the truth about Stalin's rule of terror became known. But many things distinguished him from both the Communist fanatics and the right-wing "Commie-hunters" of his era. Chief among these were, first, his complete lack of domineering zeal, and, second, his sense of humour. Even his well-known fanaticism about the North was never of the strong-arm type. Fanaticism, as the philosopher Santayana described it, "consists in re-doubling your efforts when you have forgotten your aim." Stefansson was a balanced and open-minded thinker, and the antics of political extremists were alien to him, and often best dealt, he thought, by a healthy dose of laughter.

Stefansson found it comical, for example, that he should be suspected of Communist sympathies just because he helped out his old friend Owen Lattimore by selling his farm for him — unknowingly to a man who had once been a member of the party. Lattimore was a distinguished scholar of Chinese and Mongolian studies, and as such, notoriously came under Senator Joseph McCarthy's very loose gun during the time of the rise of Mao Tse-tung.

It was also hilarious that in the January, 1948, issue of *New York Journal American*, a rabble-rousing Hearst medium, Stefansson was identified as "a veteran joiner of Red Fascist groups," and "a member of seventy-six Communist front organisations." The public was duly warned that the great explorer had undertaken the dastardly task of indoctrinating some Boy Scout supervisors in an upcoming seminar scheduled to be given at his farm in rural Vermont!

It was also rather amusing that Evelyn Stefansson, who sometimes described herself as "a Jewish girl from Brooklyn," was targeted as a "high-ranking member of the Communist party from Hungary." This was the work of the New Hampshire public prosecutor, Louie Wyman, also known as "Louie the Fox," a rabid fox indeed when it came to "Reds." He called Stefansson to appear before his local committee of investigation, and Stefansson admitted some of his memberships in suspect organizations, but defended his 1949 criticism of the activities of the House Un-American Activities Committee. Wyman neither charged him nor exonerated him (Hunt 263). To celebrate this absurdity Stefansson wrote a little poem.

> Simple Simon met a Wyman
> Trying to be fair.
> Said Simple Simon to the Wyman
> "How's your old red scare?"
> Then said the Wyman unto Simon
> "I haven't caught so many."
> Said Simple Simon to the Wyman.
> "Indeed, you haven't any."

As Gísli Pálsson notes, Stefansson had first been tracked by the FBI at the time of the Wrangel Island debacle in 1922 — and the Bureau's internal memo Pálsson quotes from that occasion is also rather full of (unintentional) humour (Pálsson 265). In the early 1950s, when Senator McCarthy asked the Navy for some information on Stefansson's work for them, they buckled under at once and assured the senator that all their doors were shut to Stefansson. In August 1951, the Senate Subcommittee on Internal Security was told by Louis Budenz, that Stefansson was probably a Communist. Budenz, a former Soviet spy who had become a born-again Roman Catholic, had previously denounced Owen Lattimore, almost certainly incorrectly, as a member of a Communist cell within the Institute of Pacific Relations, and probably knew that Stefansson, too, had some ties with that organization. Stefansson also made fun of this. "What an honour!" he said. "I would hate to be called a Communist by anyone else, but I don't mind it much from Budenz. If General Marshall and Mrs. Roosevelt can be listed as Communists ... I don't know why I can't be called one" (Pálsson 267).

Despite our retrospective sense of "the Red Scare" as a Feast of Absurdities, it must be remembered that the injury suffered by those who ran afoul of the inquisitors was very real: the misguided zealots of that era often inflicted fatal or near-fatal damage on many blameless lives and careers. Stefansson was fortunate in being able to survive almost unscathed, though even he suffered some setbacks. Perhaps his habit of speaking out, and his reputation as a wild card, as a flag-bearer for his own, mostly non-political causes, shielded him.

10

An Explorer's Passion

Stefansson and Evelyn had moved to Hanover, New Hampshire in 1953, the year he became formally connected with Dartmouth College. He had lectured at Dartmouth as early as 1929 but when he joined the faculty his inaugural lectures were especially well received, and he soon established some rich collegial relationships there. His great collection of books and manuscripts on the Arctic were purchased, as I have noted, by the Baker Library, and Evelyn worked in the library, took "courses in painting and other subjects at the university, performed with a theatrical group, and did some writing" (Hunt 262). The present Institute of Arctic Studies, established at Dartmouth University in 1989, takes as its motto a quotation from Stefansson's *The Northward Course of Empire*: "There is no northern boundary beyond which productive enterprise cannot go until North meets North on the opposite shore of the Arctic Ocean."

Stefansson and his wife, Evelyn, showing off their personalized licence plate, "IGLU."

Stefansson's last years were creative, relatively peaceful, and marked by a continuous devotion to the interests that had occupied him since his youth. As William Hunt points out, the explorer had no hobbies, played no sports, and took part in no "frivolous" leisure activities. He had always been a great reader, and reading and writing, and socializing with his friends, occupied most of his time. He had some interest in theatre, but almost none in classical music, and his appreciation of Inuit life and character never quite included the enthusiasm for its traditional and modern art that is now so much a part of our sensibility today, at least in Canada.

One issue alone may have given the old man cause for sorrow. That was his relationship to his "abandoned" Inuit family. According to Gísli Pálsson there is some evidence that this lingering, unresolved, and virtually unacknowledged connection haunted Stefansson's last years (Pálsson 268–73).

In 1955, a little girl in Inuvik named Georgina, as part of a Grade Five project, wrote a letter to her grandfather. Part of her letter read "although I have never met him I have always been proud of him, and feel that I do know him because of the stories my dad has told me about him" (Pálsson 272). There is no evidence that Grandfather Stefansson ever got this letter, which the teacher appears to have mailed, but it was published in 1961 in the journal *North* and he may have read it there the very year he had the first of the strokes that would eventually kill him. Mary Fellowes, Stefansson's secretary at Dartmouth for several years, according to information given Pálsson, had witnessed Stefansson destroying documents concerning his son Alex, and he may have confided in her, as he did in his wife Evelyn, although they had agreed not to mention it publicly. I also have the notion, and it is nothing more, that Stefansson may have told Fanny Hurst about this, perhaps in some private moment of passion, and sworn her to secrecy. It would hardly be unusual for lovers to share this kind of revelation in an intimate moment. Then too, as I have previously indicated, rumours of Stefansson's Inuit connections were rife at the time of the Canadian Arctic Expedition, and when such matters are "in the air" they often drift back, even across the years, to those involved. Stefansson's "silence" on the matter, however, was carefully maintained during his later years, and it is interesting to know that it seemed about to be broken just before his death.

In 1961, when he was eighty-two years old, Stefansson suffered a stroke that paralyzed his left leg from the hip down. He recuperated, and continued working on his autobiography, but in August of the following year, while he was enjoying himself at a dinner party with his wife and friends, he had a second stroke and within a week had passed away.

Lawrence McKinley Gould, a distinguished scientist who had worked in Antarctica, and who had been greatly influenced by a lecture Stefansson had given at the University of Michigan forty years previously, delivered the Stefansson Memorial Lecture at Dartmouth in 1962. In it, he quoted Apsley Cherry-Garrard, the chronicler of the Scott expedition, to the effect that "exploration is but the physical expression of the intellectual passion." In Gould's shrewd judgment, Stefansson was a "perfect fulfilment" of this definition (Hunt 268–69).

Vilhjalmur Stefansson travelled the North almost continuously between 1906 and 1918 and in various written accounts fashioned a most appealing vision of what was *terra incognita* to most of his readers. His work charmed and intrigued many readers, even those like Stephen Leacock, who preferred to experience the Arctic with a hot toddy in hand beside his home fireplace. Stefansson's life was an amazing personal journey from poverty and obscurity to fame and high repute. His ideas were multifarious, and often controversial, but they retain the power to challenge us. His reputation as an explorer is secure, and his anthropological work, often ignored or demeaned in the past, is being re-evaluated. But perhaps the man's greatest achievement, at least for Canadians, was his expression of the conviction that mutual respect between north and south is the proper condition for building an even more successful nation. This, at least, is the notion advanced by the former governor general, Adrienne Clarkson in her foreword to Gísli Pálsson's book on Stefansson, and it is easy to agree. Given Canada's history, and the experience of its Native peoples up until the present, it would be naïve to suggest that such a condition of harmony has been achieved. But at least the ideal has been set forth, and if the future brings something better, then Stefansson will be remembered, not merely as "the prophet of the North," but as the prophet of the *whole* North, strong and free.

Peary, Stefansson, and Greely with three members of the National Geographic Society.

11

Stefansson as Explorer

Exploring, venturing out beyond the boundaries of the known world, is an ancient human impulse, a basic drive of *Homo sapiens*, shared possibly by some of our hominid forerunners, and carried over into civilized society. As a species we are gifted with upright posture, the ability to walk on our hind legs, and we have binocular stereoscopic colour vision, which allows us to see the world around us, and to look at the stars.

Our ancestors roamed widely; they learned to observe and evaluate the places they inhabited, and to find new ones when their survival depended upon it. What might lie beyond a nearby mountain range, or across a swelling river? Perhaps a new hunting ground or pasture, or possibly something more mysterious, something wonderful or fearful, a sight to stir the soul. No doubt they knew well enough when to seek out "greener pastures" and to move on when the scene around them looked unpromising. From the hot African savannahs they spread across Asia and through Europe. During all of prehistory, such human migrations continued, fostering the development of skills that nourished further mobility and geographical adaptation.

Records of the earliest literate societies confirm the enduring quality of the human impulse to discover what wonders (or terrors) might lie beyond the boundaries of known territory. Those who led the way to the new land, or resources, often became tribal exemplars, culture heroes, yet even the most mythical of them surely reflected the adventures of very real (but unknown) flesh and blood explorers. Gilgamesh, Moses, Jason, Hanno, Himilco, Pytheas, St. Brendan, Erik the Red, Zheng He, Marco Polo — the record moves from myth and legend to actual recorded journeys that brought geographically separate societies closer together, and created an increasing sense of the vastness and diversity of the planet. Human knowledge, skills, and techniques of survival were thereby served.

In his best-selling book *Great Adventures and Explorations,* Stefansson himself discusses the claims of some of the historic heroes of discovery. In concise form he recounts how each of the more remote geographical regions was gradually illuminated by the daring work of pioneer adventurers. With his usual tolerance and practicality, he makes few moral judgments about his array of explorers. He writes: "The self-confessed greed for riches, lust for conquest, bigotry in religion, which appear through the firsthand narratives of our book, show themselves to have been powerful factors in European man's spread throughout the world. But there have been many exceptions." Stefansson cites Erik the Red, and "the Yorkshire farmer," James Cook, the British discoverer of Australia, as notable examples of "sturdy men of vision."

In fact, his favourites among explorers included the amazing Pytheas, a Greek from Marseilles (Massilia), who around 330 B.C. sailed north and west from the Mediterranean, perhaps as far as Iceland; also the aforementioned Erik, whose remarkable voyages to Greenland and beyond are equally impressive. Stefansson's third favourite was his contemporary, Fridtjof Nansen, the great Norwegian Arctic explorer and humanitarian. While Stefansson understood and respected the pathos and human tragedy of such failed endeavours as the expedition of Sir John Franklin in search of the Northwest Passage, or Sir Robert Falcon Scott's futile attempt to be the first to reach the South Pole, his deepest admiration was reserved for those adventurers who showed not only courage, but also great skill and excellent planning, and who consequently achieved something more impressive than merely an heroic and touching failure.

Stefansson's own journeys of discovery, as we have seen, were unique and impressive. Yet in retrospect they may be best understood as part of what was surely the last great age of planetary exploration, the period when the colonial empires were breaking up, and new states emerging from the dark past of empire, the period from about 1880 to 1940.

These European colonies had been created during the earlier Age of Exploration, which encompassed the end of the medieval world and the beginning of the Renaissance, and lasted from about the fifteenth to the nineteenth century. Trade was the catalyst of the new expansion. A search for valuable commodities (gold, silver, spices, precious fabrics), and the routes by which to obtain them, resulted very often in the conquest and occupation of the areas where these goods were found. Early on, the difficult land routes were by-passed and exploration by ship became the central focus. Portugal, Spain, England, France, and the Netherlands led the way in acquiring overseas colonies, to be joined later by Germany and Italy.

The colonies established around the globe by these nations were never models of enlightenment, and contradicted the avowed idea of progress that dominated European thought in the nineteenth century. The writer Joseph Conrad hardly exaggerated when he showed one of his characters (Marlow, in the famous story "Heart of Darkness") reflecting on this process: "The conquest of the earth, which mostly means the taking it away from those who have a different complexion or slightly flatter noses than ourselves, is not a pretty thing when you look into it too much…. It was just robbery with violence, aggravated murder on a great scale, and men going at it blind — as is very proper for those who tackle a darkness."

In the "twilight" period of colonialism, the era in which Stefansson went to the North, there were a few places on the globe attractive to explorers, former colonies or remote areas, still relatively untouched by development. What drew Stefansson to the North is seemingly obvious — his Norse heritage — and yet there is something of an accidental quality to his choice. (He had considered joining an expedition to Africa). Here too, Joseph Conrad's rather prescient story offers us some insight. As the narrator Marlow relates: "At that time there were many blank spaces on the earth, and when I saw one that looked particularly inviting on a map (but they all look that) I would put my finger on it and say, 'When I grow up I will go there.' The North Pole was one of those places, I remember. Well, I haven't been there yet, and shall not try now. The glamour's off. Other places were scattered about the Equator, and in every sort of latitude, all over the two hemispheres. I have been in some of them, and … well, we won't talk about that. But there was one yet — the biggest, the most blank, so to speak — that I had a hankering after."

Stefansson did not consider "the glamour off" the North. To him it was indeed "the biggest and the most blank" area on the globe, or at least the one that attracted him the most. Yet we must keep in mind that, apart from the doughty crews of the various space probes, there have been no post-modern, no post-colonial explorers. Travellers, yes, and those devoted to extreme sports and self-testing, but true explorers of our globe, not one. Stefansson, with all his baggage, must be evaluated alongside his contemporaries, or near contemporaries, those who searched out "the blank spaces of the earth". Although these later explorers were not part of the heyday of colonialism, none of them had motives that were purely idealistic; they sought fame, knowledge, power, historical connections, a glimpse of lost or vanishing worlds. They strove, very often, to preserve records of what they experienced, to ensure, by means of the conjuring acts of language, that others would share their vision, that the blank places in our knowledge would be filled.

For in reality, as they all knew, or learned at last, there were no "blanks." There was only the ignorance of the first conquerors, who in their rush to take possession of the rare and the valuable, had often overlooked the obvious and the real. These later explorers had little to do with "civilising the savages," opening trade routes, or spying for military operations, as some of their predecessors had done. They were mostly visionaries, pursuers of knowledge, and very often trenchant but eccentric thinkers, eager to prove a point, and determined to know more about some neglected corner of the Earth and/or its inhabitants. And their achievements were as various as their origins and chosen fields of exploration.

Richard E. Byrd (1888–1957) was born in Virginia into a well-known family with connections dating back to colonial times. His most spectacular achievement, for which he received the Congressional Medal of Honour, was his flight with Floyd Bennett over the North Pole in 1926, the first of its kind, although it has since been doubted that he and Bennett actually reached the pole. More substantial was Byrd's work in Little America in the Antarctic. Here he commanded an American operation that did much valuable scientific work, and here he secured his fame by flying over the South Pole, and by surviving a lonely sojourn in the darkness of the polar night, an experience he memorably recounts in the autobiographical narrative *Alone*. Like Stefansson, Byrd had great faith in science, but unlike the former, he carried out meticulous planning, (although his adventure in the polar night was nearly disastrous). As a person he was tough-minded, but also rather remote — unlike the more genial "Stef." His accounts of his adventures in Antarctica take strikingly different forms: the books describing life at Little America are factual, rather impersonal, but engaging. *Alone*, on the other hand, is almost a spiritual confession — a description of the self forced to confront the immense and impersonal cosmos, and the possibility of death.

Except for a few instances in his notebooks, Stefansson never ventured into such narrative waters; his air of objectivity, his persona of confident mastery never failed. He refused to reveal himself, or to confess to any weakness. He was not humourless, as Byrd seemed to be, but never exposed his passions, fears, or doubts. When he was attacked, or questioned, he maintained his confident air and responded always with moderation, even to the savage, half-baked critique of his views on the North by Roald Amundsen. Like Byrd, he suffered from the pressures of isolation and deprivation, but he alluded to them only indirectly. Romantic exaltation and rhapsodic celebration were alien to him; he did not indulge in any speculation remotely close to Byrd's tentative existential questionings. Gísli Pálsson, one of Stefansson's best biographers, and the editor of his notebooks, points out that the only passage in them that betrays really strong emotion describes the death of a favourite dog.

Another contemporary of Stefansson was the renowned Arabian and African explorer Sir Wilfred Thesiger (1910–2003). Thesiger was born into an aristocratic British family of Anglo-Irish origin. His father, a handsome and accomplished diplomat, died when his son was ten years old, which only strengthened the boy's lifelong attachment to his remarkable mother. Thesiger, who attended Eton and Oxford, and who had been physically punished and sexually abused by a sadistic preparatory school master, soon became devoted to physical discipline; he neither drank nor smoked, and had, he claimed, "no interest" in heterosexual connections — his emotional attachments were exclusively to men. He took many photographs of his young Arabian and African companions, "adopted" some native boys and took them to live with him, and late in his life wrote a private monograph recording in precise detail the male circumcision practices of the Samburu people of Kenya.

Despite a certain guardedness in their natures, both Thesiger and Stefansson had lively friendships across a fairly broad spectrum, and greatly valued these. Yet in contrast to Thesiger's romantic and very poetic male-to-male relationships (which were no doubt also sexual in some cases), Stefansson found a very down-to-earth companion, a "native wife," Pannigabluk, an Inuit seamstress, who bore him a son. Like Thesiger, Stefansson was comfortable in two worlds: while making his name as explorer and hunter, and maintaining a deep sympathy with the Native peoples he encountered, he also functioned very comfortably among artists, writers, and other sophisticates in urban society. Stefansson chose his various lovers from among New York career women, including, as we have seen, the well-known novelist Fanny Hurst.

Thesiger's expeditions had little to do with science, although his books, written with far more literary panache than Stefansson's, increased the knowledge of his readers and gave them strong and accurate insights into the life of vanishing tribal peoples and threatened landscapes. Thesiger was, almost unwittingly, one of the main recorders of the twilight of colonialism. Directly connected to the imperial system through his family, he mostly circumvented any direct criticism of British colonial rule. His vivid evocation of various local people, strongly foregrounded in all his books, might even be viewed as a successful strategy of avoidance of such criticism, although it is also clearly a celebration of individual human worth, of empathy and love felt across cultural boundaries, and the insidious divisions that usually exist between rulers and the ruled. Stefansson did not manage to evoke the people he encountered with similar literary skill; but in some of his stories and observations he is effectively satirical about certain groups, notably missionaries, whose work he often exposed as inept and misapplied.

Thesiger could be described as an enemy of "progress" as it is understood in Western Europe and the United States. He hated machinery, while Stefansson, although often careless about maintaining the machinery he relied on, required it to further his projects. (Yet he also occasionally mocked "development" in the North wherever it undermined centuries-old adaptation by the Inuit peoples.)

Both of these physically impressive and courageous men courted fame in a similar manner — both built their careers in a way that was steady and insistent, rather than showy and obvious. Although he often travelled with companions, Thesiger was at heart solitary, while Stefansson, despite his penchant for occasionally going it alone, had to accept serious responsibilities as the leader of his expeditions. This led him into controversy, a controversy quite different, and ultimately more serious, than those that plagued those other expedition leaders, Peary and Byrd. While most of Thesiger's explorations were carried out as part of the last gasp of the dwindling British Empire, his connection with that world was transparent and relatively blame-free, while Peary's invocation of American patriotism to further his personal ends, together with the possible spuriousness of his claim of a polar "first," seriously undermines his self-created heroic image. Byrd's fabrication of his North Pole "first flight"— and his subsequent acceptance of the Congressional Medal of Honour — turns him into a bit of a charlatan. There is a touch of charlatanism, too, in Stefansson's blithe brushing off of the catastrophes associated with his third polar expedition and the Wrangel Island adventure, since he pretended that those real catastrophes resulting in loss of life and great suffering for the survivors were simply "accidents" over which he had no control. Unlike Peary and Byrd, however, Stefansson was made to suffer during his lifetime for his misjudgments, and although he has been taken to task by later writers for some of his actions as expedition leader, the fact that he carried on, and fashioned a distinguished and productive life to the very end, should count in his favour. After all, it's just possible that he really believed that he had sufficiently fulfilled his responsibilities as a leader, that it was the fault of others that they had succumbed to the trials and dangers of "the friendly Arctic."

Stefansson, like Thesiger, and like certain other explorers — Nicholas Roerich, (1874–1947), for example, the Russian artist who travelled in central Asia, and Laurens van der Post (1906–1996), the South African novelist who wrote of the Bushmen of the Kalahari Desert — in various ways sought to "enter" the indigenous society he encountered. As I have several times noted, Stefansson mastered Inuit survival skills, and, while on the trail, lived as a Native, adopting local dress, mastering hunting and fishing, subsisting on the typical Native diet, and making use of dogsleds for transports. His attempt to correct the stereotypical image of the Inuit held by the white society of his time is not unlike van der Post's "corrective vision" of the Bushmen and their seemingly harsh and minimal way of life. But while van der Post and Roerich sought "deeper mysteries" in Africa and Asia, mysteries that they saw as useful in subverting Western materialism and rationality, Stefansson's attitude to Inuit beliefs remained objective and skeptical. Van der Post associated the Bushmen and their environment with Jungian archetypal experience, and asked his readers to consult their own psychic life to gain an appreciation of the "first things" manifested in the Bushman life and culture. Roerich was obsessed with the ancient Slavic culture, and it is no accident that he provided the basic scenario and some of the costuming for Igor Stravinsky's famous musical venture into "the primitive," *Le Sacre du Printemps.* In his Asian expedition of 1925–1928, Roerich sought Shambhala, the mystical centre of theosophical teachings. In later years he settled in Himalayan India, and continued to espouse and illustrate esoteric Buddhism. Stefansson, on the other hand, delighted in exposing what he saw as the naiveté of Inuit beliefs, and the "simplicity" of a people whose most significant acquisition from Christianity was a dogged belief that no work should be performed on Sunday, even when the whole community's survival was at stake. He presented himself as an earthy, shrewd, and sociable descendent of Vikings, a man of this world, strongly grounded in the everyday, with a strong belief in the power of science to explain and transform reality.

Gísli Pálsson contrasts Stefansson's amalgam of exploration and anthropology with modern approaches:

> For him, and many of his contemporaries, fieldwork and geographical explorations were, above all, exercises for testing and strengthening the sensibilities of manhood against all odds. Accounts of such gallant journeys inevitably place the Natives in the back seat whatever their real contributions. It would be silly to force modern methodological standards upon Stefansson's approach and viewpoints. The point is not to establish that he failed to conform to our standards, which seems rather obvious, but rather to explore the differences between the two contexts. Most anthropologists nowadays would argue that it is important to be explicit about the effects of one's presence on the scene as well as in the ethnographic texts. As a result, anthropologists generally feel compelled to situate their accounts and to reflect on the texts they write as well as their relations with their hosts and readers. (Arctic Web 2)

In other words, Stefansson was one of the last of the traditional explorers and anthropologists, with all the limitations that implies in terms of methodology, but in the light of his amazing journeys, he was also a forerunner of our contemporary devotees of "extreme sport" and self-development through physical testing. His admiration for Erik the Red and others who have been credited with amazing journeys is not surprising, given his own astonishing feats of exploration. He travelled, with few modern conveniences and with none of our contemporary technological adjuncts, some 32,000 kilometres on foot, and explored some 160,000 square kilometres of Arctic territory. The land he traversed — despite his various reassurances — was bleak and dangerous. He had to face fierce storms, ice and snow, isolation, darkness, and the constant threat of starvation. The sheer effort required to establish where he was on the trail, to find shelter, eat, stay warm, and at the same time gather information about the land and peoples he encountered, must have been prodigious. Only a few of his contemporaries come close to his achievements in terms of bravery and boldness in relation to such daunting challenges.

A comparison of Stefansson with his near-contemporaries in the field of polar exploration may be instructive. While he professed to admire Amundsen for his much-lauded careful planning and sensible adaptation to local conditions — some would say, "ruthless efficiency" — Stefansson himself seemed almost incapable of this kind of precision. He was a man of impulse and improvisation, and his reputation suffered from his failure to respect what we have since come to know as "Murphy's Law," namely, that "if anything can go wrong it will." His habit of assuming that nearly all his colleagues were as resourceful in the field as he was may have been at the root of this problem. As for Amundsen's famous attack (in his 1927 autobiography) on Stefansson's notion of "the friendly Arctic," Stefansson claimed, in a later edition of his own famous book of that title, that the Norwegian's ire had actually promoted controversy and discussion, and therefore improved sales. Since it appeared that Amundsen had not actually read *The Friendly Arctic*, but was speaking on the basis of what he had heard about it, his criticism did not really upset Stefansson, although the latter never deceived himself about Amundsen's disapproval of his characterization of the northern environment.

Stefansson also admired Knud Rasmussen, the great Greenland explorer, and resembles him in some respects more than he does most of the others. Like Rasmussen, he was a good linguist, and felt that understanding the Inuit languages was necessary in order to understand their culture, an idea that seems obvious now, but was not universally respected in the "imperial" age of exploration. Stefansson and Rasmussen both imbedded themselves in the Inuit cultures they explored, although it's probably fair to say that Rasmussen's books have more ethnographic respectability than Stefansson's.

Stefansson bears comparison to Shackleton in that in the field both men were enormously strong, cool-headed, and courageous. They both loved and cultivated publicity and, perhaps as a result, both were considered too "forward" in their demands and had the label of "bounder" occasionally thrown at them. Neither was really such, but both, in the eyes of some, wanted to climb too fast and didn't conceal the fact.

Stefansson was a great admirer of Robert Peary, who, whether or not he actually reached the North Pole, was a courageous and impressive traveller, one who developed some of the basic techniques of Arctic travel. He, like Stefansson, fathered a son with an Inuit companion, and concealed this side of his life. To his discredit, Peary also failed to acknowledge the achievement of Matthew Henson, his African-American assistant, who reached the North Pole area with him. Nor did Peary approve of Stefansson's part in obtaining an Explorers Club citation for Henson. Stefansson's opinions about Peary were rather carefully guarded, probably because he needed the older man as an ally, and didn't want to offend him by criticism. (Peary's sensitivity to criticism was notorious.) Very likely Stefansson was suspicious of some of Peary's calculations, and doubted the accuracy of detail in his polar narrative, but he blamed the latter on Peary's "ghost writer" and his depressed state at the time.

Although Stefansson advocated the establishment of commercial air routes over the North Pole, and wrote approvingly of air exploration by the Soviets and others, he himself explored by land. Ships were a necessary evil, and flight detracted from the real experience of the Arctic explorer, he believed, though it might be a means to an end. His trail-mate George H. Wilkins, who began as a superb photographer, and ended up as a celebrated flyer, and as the proponent of the first submarine voyage to the North Pole, was in many ways Stefansson's protégé, although occasionally nervous about the master's methods. When Stefansson discovered new islands, Wilkins expressed doubts about the usefulness of such activity, but Stefansson might have asked in return whether flying over land and water adds much to human knowledge. Wilkins, like Byrd, had a fine sense of using aircraft for gaining wide vistas of unknown territory, but the machinery employed was often so primitive, and liable to failure, that not much time was left for leisurely perspectives.

Stefansson was a great explorer, one of the last, and he belongs in a select company. Taken all together, these men filled in those "blank spaces of the earth" referred to by Conrad. If some of them cheated, or sought fame to inflate starved egos, or defended their own territories with a too ruthless sense of entitlement, we can forgive them. Brave as they were, inimitable as they were, the tasks they undertook — compared with the vast global problems that confront us today, pole to pole — were merely the games of boys, lured by danger and difficulty. "Exploration is the poetry of action," wrote Stefansson. They were lucky to find that poetry, to pass along their special knowledge of a world still mysterious, one not yet hopelessly tarnished by human carelessness and greed.

Stefansson (foreground) leaves the Karluk *in September 1913 with a hunting party: he had left a letter saying he would "be back to the ship in ten days" but in fact he never returned and spent the next five years exploring the northern Arctic.*

12

Stefansson and the North

Vilhjalmur Stefansson's wide-ranging investigation of the ecological, social, and cultural relevance of the Arctic, most of which was completed by the 1920s, and his many articles and books celebrating aspects of northern life, earned him the sobriquet "Prophet of the North." To those who know something of the reception of Stefansson's ideas, particularly in Canada, or have the slightest notion of the implications of global warming, this title is full of ironies.

Stefansson's book *The Northward Course of Empire* was published in 1922. It is a visionary narrative, full of grand ideas and bold predictions, a book far ahead of its time — the kind of book that causes bureaucrats and "sober scholars" to nod, shake their heads, and look the other way. Stefansson's thesis assumes that expansionism and "the conquest" of new frontiers is a near-permanent element in the history of the developed societies of the West. But there are positive and negative routes for such development. Challenge and struggle, such as faced Erik the Red and the Vikings, or the later explorers who sought the Northwest Passage, can be rewarding, for to win against great adversity is more character-building, more stimulating, than easy conquest. Stefansson's idea of historical experience is based on popular Darwinism and is inclined to favour the struggle implicit in "hard primitivism" over the *dolce far niente* of "soft primitivism." From this perspective it is not surprising that Portugal and Spain, colonizing nations that expanded south, lost out in the end to the British, who, with the French, took on the challenge of North America. Like the Danish writer and Nobel Prize winner Johannes Jensen and the British historian Arnold Toynbee, Stefansson assumes that nations that have too easy a path tend to fail the ultimate test of "development" and "greatness."

Moving, somewhat paradoxically, from this assumption, Stefansson goes on to argue that, although the northern challenge can be a severe one, the character of the North has been greatly misunderstood. Here Stefansson's notion of "the friendly Arctic" comes into play. The popular image of the North as an eternally cold, snowy, dark region hemmed in by ice and to the last degree treacherous to its inhabitants is quite false. The North is surprisingly varied in its landscape, weather, flora, and fauna. It is by no means physically or psychologically daunting in all of its aspects; and permanent settlement is no fearsome prospect. On the contrary, the Arctic can be fruitful in many ways. Natural resources, oil and mineral wealth, will in the future form the basis of lucrative industries there; cities will grow up; transportation will improve; and new air routes will link the northern settlements to Europe and Asia. New food sources deriving from northern animals and fish will become commonplace in the North American diet. Earl

Hanson, one of Stefansson's biographers, reports on a lecture that he attended, in which the explorer "opened the gates for us to a new North ... a vast, friendly, and fertile region where men would someday herd reindeer by the millions, build cities, mine coal and copper and gold, run health resorts on the shore of the Arctic Sea, and send their children to school."

Although Stefansson's ideas were cordially received in some quarters, and his reputation grew in England and the United States, in Canada he remained for a long time a prophet whose visionary zeal and particular expertise were ignored. The Canadian government of the day found his call to discover and develop the North rather irrelevant to its political aims, the general public was not inspired, while the lingering resentments of a few colleagues who had been prominent on the Arctic Expedition of 1913–1918 worked against him in the heart of the Canadian bureaucracy. The Liberal government of Mackenzie King, which succeeded that of the Conservative and war cabinets of Sir Robert Borden and Sir Arthur Meighen, shied away from the "controversial" explorer, and ignored his theories about northern development. The Wrangel Island disaster and the failure of the Stefansson plan to domesticate the reindeer on Baffin Island in the early 1920s did nothing to help Stefansson's cause.

For the time being, most Canadians continued to regard the North as a frozen wilderness, and the northern one-third of the country was left to "the missionary, the fur trader, the Eskimos and the Indians," as a Canadian minister of northern affairs put it, as late as 1957. This was not surprising, since Canadians had traditionally settled and built their country in the relatively salubrious geographical corridor that ran along the United States border. Yet, after the Second World War had elevated both the United States and the Soviet Union to positions of world power, and those nations began to take note of their own northern potential, some Canadians, too, began to look northward.

The Soviets had demonstrated the viability of settlements in bleak Siberia, begun the Arctic flight patterns espoused by Stefansson, and laid claim to vast areas of the North. The United States, thanks to its huge foothold in Alaska, and based on the explorations by Frederick Cook, Peary, and others, was beginning to look north. It did so for both strategic and economic reasons, as a natural response to the perceived Soviet threats, and as part of its own historic expansionism. In connection with the many defence postures directed at the Soviets, including the Distant Early Warning system (a set of bases established by the United States across Canada to monitor a possible Soviet air attack), numerous American military personnel appeared in the North. At the same time, American and Soviet submarines patrolled Arctic waters. In the light of such developments, Canada, despite having no large global claims, slowly but inexorably came to see not only the need to assert its own sovereignty, but also began to conceive of the North as a logical area for national expansion, a difficult but not intractable *Lebensraum* where, in the near future, the most striking economic and social development might take place.

In 1953, the Canadian government of Louis St. Laurent finally created a Department of Northern Affairs and National Resources. (The link between northern territory and the "potential wealth in the ground" is significant.) And the Conservative government of Prime Minister John Diefenbaker in its 1957–58 election campaign promoted a new "vision of the North," one closely allied to what Stefansson had put forth some thirty-five years earlier. According to historian Desmond Morton, "Diefenbaker's 'vision' had been conceived by an advertiser as a development plan for Canada's northern frontier: it became a catch-all for miscellaneous but useful projects like the South Saskatchewan Dam, the railway to lead and zinc deposits at Pine Point on Great Slave Lake, or the development of Frobisher Bay on Baffin Island" (Morton 224). Perhaps not exactly what Stefansson had in mind, yet Diefenbaker later went out of his way to praise Stefansson quite specifically, and commented: "He experienced many discouragements in his efforts to make the North live in the hearts and minds of Canadians" (LeBourdais 186).

Stefansson himself was well aware of some of the ironies of the situation. For one thing, his experience told him how successfully the Native cultures had adapted to the harsh conditions of the North, and he occasionally noted how superficial attempts to change the relation of the Inuit to the land were unlikely to bring real improvements. Often, he wrote and spoke as if he were just one more apostle of unthinking "progress." Yet neither his romantic sense of the value of struggle and individual effort, nor his understanding of the viability of traditional beliefs and customs, encouraged him to embrace without question such progress when it appeared in the naïve form of development at all costs. Even more important — and unlike many who advocated northern development — Stefansson was both intimately familiar with the conditions of northern living and well informed about its environment, history, and mythology. True, Stefansson's advocacy of the Inuit seemed to focus on issues that we would regard as extremely marginal, or merely faddish: their diet, for example; while his analysis of their stories and ritual practices seems more ironical than sympathetic. Despite this, he better than most understood that many historic northern adventures had ended in disappointment and tragedy, and that blithe visions of a new frontier must be tempered by an awareness of reality.

In short, despite his advocacy of "progress" and expansion,

Stefansson had a more complex idea of the North than those who saw it chiefly as a sphere for the development of mines and oil fields, or as a problematic region of "native affairs."

Here a few clarifications of terminology seem called for, together with a few notions of what our Western history, myth, and stories reveal about the North, some of which formed part of Stefansson's vision.

As is clear from the above, the "North" is a much broader concept than the "Arctic." Yet both words are often used rather loosely, and remain vague, hazy, undefined in any rigorous sense. Both words carry an air of mystery and breed mythologies. Understandably enough, since when one calls to mind any one of several characteristic "northern" scenes, a sense of "liminality" prevails: boundaries are blurred and indistinct, and much in the landscape can be seen to subtly shift, move, or change. When we think of the North, or the Arctic, we see ice drifting in open water, boundless realms of packed snow, forests shrouding the bedrock of mountains, tundra without borders or definition. We picture melting ice, glaciers in slow, near invisible movement, evergreens swallowed in mist, icebergs mostly hidden in ocean, blizzards concealing the contours of the land.

The "North" mostly refers to those parts of Europe, Asia, and America farthest from the equatorial regions, areas covered deep in ice in the geologic past, the lands from which the great glaciers retreated, leaving boreal forests, bleak mountains, bare, boulder-strewn plains, and deep cold lakes. Mythologically, this is the region of the often phlegmatic, but sometimes violent and energetic "Nordic" races, the home of the shamans, the Norse gods, and, in song and story, the regions of fire and ice.

Somewhat less mystically, the "Arctic" is often defined as the area beyond tree growth, the land beyond the Arctic Circle at

66° 30', an area that includes the northern parts of the mainland of Canada, Russia and Alaska, Greenland and the smaller polar islands, such as Spitzbergen, Ellesmere, Novaya Zemlya, Melville, and Wrangel Island, among many others.

The North has often been the *Ultima Thule* of the imagination, the Arctic itself the vast, domed "crown" of the Earth, which as children we think of as that place "up there," the region directly above us, the land of the North Pole, or Santa Claus land; also — in less fanciful, though still poetic terms — as the round, rather remote "ceiling" of our beloved blue planet. The north and south poles are often lumped together as icy wastelands, but in truth, the "top" of the Earth is very different from its southern polar counterpart, even though they have a few fundamental features in common. Certainly, both regions are very cold, with abundant snow and ice. Both have long, dark, sunless winters and perpetual daylight in summer, thus confirming their status as regions exhibiting these primal "polar" opposites. The differences, however, are substantial. Whereas the Antarctic is a continent surrounded by seemingly endless ocean, the Arctic is an ocean surrounded by vast continents. This ocean features a wide continental shelf, relatively low salinity, and two-thirds of it is covered by drifting pack ice. Marine life is rich, and large mammals (polar bears, seals) are native there. Most notably, and in contrast to its southern equivalent, the North — in some areas — has long supported thriving human cultures. Antarctica, by contrast, is a truly frigid continent: the Antarctic is colder on average than the Arctic wastes; it features high elevations, mostly covered by permanent ice caps which form shelves (recently noticed to have been melting, due to global warming). These shelves border the Antarctic for at least half its length, and give the region the aspect of a great natural bastion, whereas the Arctic is more diffuse and amorphous in character.

The differences in geology, climate, and landforms between the two poles are reflected both in their mythologies and in the history of exploration associated with them. The South Polar Region — remoter and harder to reach — was fancifully populated with "antipodal" humanity, and became a favourite area in which to locate Utopian and science fiction fantasies. The poet Coleridge immortalized it (and its most spectacular bird species, the albatross) in his poem *The Rime of the Ancient Mariner*. Edgar Allan Poe, possibly influenced by theories of underworld access via the poles, imaged it as the gateway from our "grotesque" reality to another world, the "arabesque" world of spiritual perfection. John Taine and H.P. Lovecraft wrote about stupendous and rather frightening extraterrestrial races that had supposedly settled there.

The stories and myths of the North, as we shall see, are even more detailed and trenchant in their relevance to the fate of the human species on this planet. This is to be expected since the contact of important Western societies with the northern regions and peoples has been so extensive and long-standing.

Yet despite such contact, there is often an uncertainty about the history of northern exploration. Do the often opaque or vaguely defined northern landscape and its murky weather play a part in this? Indeed, the Northern Lights, perhaps the most poetic and spectacular of natural phenomena, symbolize a very important aspect of the northern imagination, being incredibly beautiful, and presenting an aura of mystery, despite having an origin explained by science. However we look at it, the mysterious-in-the-natural is certainly an important aspect of the northern regions of our planet.

Stefansson himself published a well-researched and fascinating book called *Unsolved Mysteries of the Arctic* (1931). And indeed, when we think of the explorers of the North, including Stefansson, we see that their achievements are often unclear, that mysteries abound, that things are shrouded in a fog of uncertainty and controversy. Stefansson's essays, all of them rational probes in his characteristic manner, testify to the elusiveness and ambiguity surrounding the exploration of the North, and decades of scientific work and historical research have not quite dispelled the image of a region marked by "liminality" and uncertainty.

Consider the medieval Norse settlements in Greenland, now vanished, which have been closely studied, and whose demise has occasioned so much speculation. These Norse settlements were much more than short-lived base camps or trading posts; in fact, they flourished for nearly five centuries. Scholars are still arguing why they disappeared. Was it disease, a failure of adaptation, attacks by the Inuit, or by outsiders, or some other factor that spelled the end?

Among many theories, Stefansson suggested that the settlers might have intermarried with the Inuit, an idea unproven until recent genetic testing was undertaken. As Gísli Pálsson notes, such testing now confirms that Norse men had children with Inuit women, thus establishing that the Greenland Colony was never overrun or conquered by the Inuit. (If it had been, one would expect Norse women to have had children with Inuit men.) Some mystery still surrounds the fate of the Greenland Colony, however, and Stefansson was quite right to see it as an enduring historical puzzle.

Much has been written about the catastrophe of the 1845 Franklin Expedition. Few failed voyages have resulted in so many inquiries and follow-ups. Here, bad planning, possible food or lead poisoning, and probably cannibalism, not to mention incompetence in Arctic travel, buttressed by notions of racial superiority, make their unfortunate appearances.

Who discovered the North Pole? Was it Peary in 1909 or Cook in 1908? Or neither? Did Admiral Byrd, that great American hero, really fly over the pole in 1926, as he claimed? (Some deny it, and he is said to have privately admitted his fraud.) Why did Stefansson abandon his ship the *Karluk* during the Canadian Arctic Expedition of 1913–1918? Was the later gunshot death of one of the *Karluk* crew murder or suicide?

South Pole exploration, by comparison, seems clear-cut. The Antarctic gave us the tragedy of Sir Robert Falcon Scott, the dogged heroism of Sir Ernest Shackleton, the ruthless efficiency of Roald Amundsen — their famous journeys were heroic and/or tragic, but relatively straightforward. Not so in the North, where even the shrewd and cautious Amundsen was to meet disaster. What happened at key moments in northern exploration is often in dispute. Nobody doubts that Amundsen attained the South Pole on December 14, 1911, or that Shackleton accomplished one of the great feats of exploration in that region, or that Richard E. Byrd spent part of the winter of 1937 alone at a weather station at 80° 8' south, 163° 57' west, on the Ross Ice Shelf. But in the North, as Stefansson argues, uncertainty about key events is fairly common.

A brief look at northern mythology and stories should help us to understand this seemingly implicit uncertainty.

We might begin with shamanism, which is so much a part of Inuit culture. This practice, although certainly not limited to the North, is very much a northern phenomenon. What is shamanism? It is nothing less than the trance journey of the tribal healer into the other world. This journey encompasses a carefully worked out ritual which includes: control of a magic fire, a meeting with helping spirits, the ascent of the world tree, a journey's end contact with the sun god or celestial partner, and a return to the original reality with some message to the individual or group.

In 1979, many years after Stefansson's death, an Inuit mythical drama was performed by a Greenland company at the Château Laurier in Ottawa, a hotel he often stayed in. The following glimpse of the performance enables us to measure something of the gap between Stefansson's sense of Inuit culture and ours:

> Reidar Nilsson, the director ... managed to explain the basic elements of the drama we were about to see. Then from somewhere among the chairs, the first actors appeared, and sealskin loincloths made their debut among the bored, the puzzled, the expectant and the uncomfortable. The drama that followed was touching and gutsy, naïve and powerful, mysterious yet comprehensible. Two youths and two girls live in happy innocence, and dance out the joy of Edenic pleasures, but are called upon to suffer when life changes and trials come upon them. They must put on the masks and robes of suffering and go through a ritual of transformation, in which, through the agency of a shaman, they encounter various terrifying spirits, and are at last redeemed and reunited in a triumphant climax marked by strong sexual tensions. Through most of this we sat riveted because of the terrifying masks, which reminded us of the closeness of the Inuit to the animal world, and called up various archetypes familiar to us through psychology — the shape-shifter, the terrible mother, and others — but there was nothing intellectualized about this theatre of physical confrontation in which clouts,

> smacks, kicks and thumps resounded through
> the room, scaring a child in the balcony into its
> own private shrieking trauma, and reminding us
> of the athletic energy of primitive art, and that a
> shaman's trance is not a polite parlour trick, but
> a wild and lunatic orgy, closer to the gyrations of
> a rock star than to the delicate manipulations of
> a Christian priest. (Henighan, "A Few Thoughts
> on the Northern Imagination.")

In 1979 there was still a great deal of condescension in our attitude to Native art. Perhaps there still is. But as more authentic material percolates into the mainstream, for example, the recent film *Atanarjuat: The Fast Runner* (2000), our shopping mall culture can experience the scary raw breath of the North, the passion, the conflicts, the bleak landscapes, the shamanic trances, and the powerful struggle between good and evil. "You never know what's out there," one of the characters in the film says, as two Inuit camp at night in some lonely Subarctic place.

These are elements of northern culture, and Inuit art, that Stefansson never conveyed to his contemporaries. Possibly, like many of his era, he narrowed his sights too much in the name of "rational analysis" to have appreciated them. Yet if he had visited the recent exhibition of contemporary Inuit sculpture exhibited in the National Gallery of Canada in 2007, he might possible have been surprised by the skill and depth with which young Inuit artists are combining traditional imagery and stories, and ideas derived from shamanism, with forms that reflect the sophistication of mainstream modern art.

In fact, shamanism, which Stefansson tended to disparage, has been quite credibly related to many of our early art forms, for example, to the myth of the hero and the epic poem featuring the journey of the hero from this world to the other and back. Nor is it surprising that scholarly articles have been written explaining how shamanism underlies many of our popular entertainments.

One can make wonderful free associations to connect shamanistic rituals with the heart of Western mythology. The shaman's magic fire and the forest trees take us close to Wagner. The shaman's drum, made out of the bark or wood of the world tree we can perhaps hear in Carl Nielsen's great Fifth Symphony. ("There is something very primitive I wanted to express," the composer tells us, "the division between dark and light, the fight between good and evil.") The world tree of Norse mythology, Yggdrasill, is surely an outgrowth of the shamanic tree. We may recall J.R.R. Tolkien's fascination with dark forests and tree spirits. D.H. Lawrence found the Pan spirit in America among the pine trees of New Mexico. The Norwegian novelist Knut Hamsun discovered Pan in the wooded mountains of the Nordland, the Far North of Norway. The symbol of the spirit-bird which permeates our literary imagery is not surprising when we think of the shaman's oft-worn coat of feathers. The talking animals of our fairy tales perhaps go back to the shaman's power to converse with nature. The masks and clowns that we know from the circus, Greek theatre, and the *commedia dell'arte* are all relatable to the primary rituals of shamanism.

The Northern Lights, shamans, spirit journeys, animal helpers, explorations that fade into mystery, all these characterize the Arctic lands. Truly, the North is a place where boundaries are ever-shifting and history is written on melting ice.

From such a perspective, the grandiose and at bottom rather mundane vision of the North espoused by John Diefenbaker in 1957–58 seems to have little to do with the northern imagination. Economic development, however essential, often destroys the nature–culture balance that creates the *genius loci*. To define the northern imagination we have to return to myth, art, and literature.

For example, if we look at the Norse pantheon of gods, and compare it with that of the Greeks, we get a sense of the roots of the northern imagination. The Norse gods and goddesses, however powerful, are always threatened. The decrees of fate, the brutality and cunning of the giants, and the violent end of the world at Ragnarok — such things circumscribe their power. As Tolkien reminds us, "The Greek gods are not besieged, nor in ever-present peril or under future doom … in Norse stories the gods are within time, doomed with their allies to death. The battle is with the monsters and the outer darkness. They gather heroes for the last defense." In that connection, it appears significant that though the mythical image of the three Fates originates in Greek mythology or earlier, it is the Norse image that we mostly remember—the three goddesses as spinners of destiny. There they sit, at the bottom of Yggdrasill, the world tree, Urd, Skuld, and Verdandi, fate, being, and necessity, spinning out the destinies of men and women. The old English word w*yrd* or "fate" is a significant one in the northern imagination.

In the great northern epic *Beowulf*, in which the word w*yrd* appears no fewer than nine times, at the moment of the hero's greatest triumph, these sombre words are spoken to him (I quote Burton Raffel's translation), "Your strength, your power, are yours. For how many years? Soon you'll return them where they came from. Sickness or a sword's edge will end them, or a grasping fire, or the flight of a spear, or surging waves, or a knife's bite, or the terror of old age, or your eyes darkening over. It will come, death comes faster than you think, and no one can flee it."

Beowulf itself, as the poet Robert Bly noted, is the very

source and mainspring of the Western imagination. It gives us a glimpse of a dangerous world, a northern world made up of (to quote the poem): "secret places, windy cliffs, wolf dens where water pours from the rocks, then runs underground, where mist steams like black clouds, and the groves of trees growing out over their lake are all covered with frozen spray, and wind down snakelike roots that reach as far as the water and help keep it dark. At night that lake burns like a torch. No one knows its bottom. No wisdom reaches such depths."

No wisdom reaches such depths ... *Beowulf* is important because it sets the task for our culture of reconciling two opposites: the dark labyrinthine underworld of monstrous forms, and the aspirations embodied in the golden-roofed Herot, the palace built by Hrothgar the Dane. The upper world is made safe by the heroic quest of Beowulf into the darkness. If our civilization is to survive, wisdom must acknowledge, penetrate, and transform an almost unfathomable evil. Those who naively embraced northern "development" have not learned the messages of our most profound northern mythologies. It is certain that Stefansson, who read the Norse sagas and myths at home, was familiar with these.

Beowulf originates in the eighth century of Western culture. It emerges just previous to the time that the Vikings began their remarkable voyages that touched so many parts of the northern hemisphere. The northern imagination owes much to the Vikings, the Norse peoples. Their mythology expresses the northern imagination, as does their poetry. And in the Icelandic sagas we have some of the greatest storytelling in prose ever achieved. The sagas were written from the middle of the twelfth to the beginning of the fifteenth century. The word *saga* means story, but these are stories in which imaginative fiction is rooted in biography and family history. These tales are foregrounded in action, with little depth of perspective, and they are closely connected with kin and family traditions and with real places. What makes them a central part of the northern imagination, however, is their power of combining matter-of-fact reporting with the spookier aspects of reality. The sagas are not mythological, but they are full of magic and the supernatural.

What connects Norse mythology, the old English poem *Beowulf*, and the sagas, is not only the grim northern setting, but an awareness of the harshness and the uncertainty of existence, combined with the sense that life is unpredictable, strange, and even weird, to use our modern version of the old English word.

But the poems and stories I mentioned are old stories. What about the modern world, the world that owes so much to science and progress? What happens to the northern imagination there?

After Darwin published his book *The Origin of Species* in 1859, there was a revival of interest in what one might call the northern world-view. Before Darwin, Romanticism had emphasized the ideal and uplifting qualities in nature. It's true that certain works of Romanticism had looked at the bleaker side of humanity in nature and society. Mary Shelley's *Frankenstein*, for example, ends with a vision of the Frankenstein monster disappearing in an Arctic world of "darkness and distance." But after Darwin, literature, like philosophy and sociology, began to emphasize the notion of life as a struggle for existence, and to focus on the grim and inhuman aspects of nature. The so-called Naturalist school included post-Darwinian writers who tried to introduce ideas of heredity and environment into their narratives. Yet despite the fact that almost all of the Naturalistic writers depict human life as determined by impersonal forces, they treat such concepts in the manner of the old wyrd, or doom of the gods. They also introduce luck, magic, the supernatural, and superstition into their stories, thus continuing the oldest traditions of the northern imagination.

Toward the end of the nineteenth century, the North and its harshness began to gain a favoured place in the literary imagination. In 1878, describing the bleak landscape of Egdon Heath in his novel *The Return of the Native*, Thomas Hardy affirmed that "the time seems near, if it has not actually arrived, when the chastened sublimity of a moor, a sea or a mountain will be all of nature that is absolutely in keeping with the more thinking among mankind." Sober vacationers, Hardy suggested, might even abandon the sunny southern landscapes in favour of trips to Iceland. This was going a bit far, no doubt, but Hardy was not alone in his sense that the northern imagination had something especially relevant to offer the modern consciousness. William Morris, the great Victorian poet and designer, translated the sagas, and made a notable trip to Iceland in 1871, a trip which stunned him and transformed his literary style.

The literary Naturalists, writers like Jack London, were producing stories in which the bleak northern environment became a key element. In their stories struggle was paramount, the species was more important than the individual, and primordialism, the celebration of the primitive, was rampant. Around the beginning of the last century, hard primitivism was in vogue everywhere — the idea that to become strong you have to overcome obstacles, to be tested, most often in some inhumanly harsh environment. (Such a view is the direct opposite of soft primitivism, which is reaching perfection by lying on a tropical beach: by entering the new Eden, or Margaritaville, you achieve bliss and forget all your troubles.) By contrast, hard primitivism says that you improve by meeting challenges and by enduring hardships. No pain, no gain. And although the most famous and popular works of hard primitivism, the Tarzan stories, were set in Africa, they owe a lot to the northern imagination, which sees struggle as perennial and toughness and cunning as the key to survival. When the young Stefansson chose to join a northern expedition rather than one to Africa, he opted symbolically for a life devoted to the experience of "hard primitivism." Africa, of course, might have led him to equal or even harsher realities, yet somehow his choice of the North seems appropriate, given his heritage of Norse culture and story, and our cultural sense of the North as the place where one gains strength through harsh testing. For Stefansson the North became a sphere of personal transformation, and in his writings he presented a classic answer to the negative side of "going into the harsh primitive," namely the discovery that its harshness can be tempered by experience and knowledge, and that if one learns to work with the conditions presented one may come away triumphant.

In the northern imagination, inner and outer often coalesce in images of natural bleakness or human suffering. Late-nineteenth-century Europe was in crisis. The dream of eternal progress had been undermined. Colonialism was yielding bitter fruit. Christianity was suffering the death of God. The old order was giving way to intensified class struggle, and the traditional patriarchy was being challenged by the new feminism. An increasing sense of despair and cultural exhaustion produced what one might call the "neurasthenic North," most familiarly represented by Edvard Munch's *The Scream*, by the plays of Ibsen and Strindberg, the novels of Knut Hamsun, by Hardy in *Jude the Obscure*, in Hugo Alfven's Fourth Symphony, and in the popular *Valse Triste* by Jean Sibelius — as well as later in the films of Ingmar Bergman. The outer bleakness and grim sense of life of the traditional stories was transferred inward to create the sense of a bleakly oppressed psyche, unable to believe in God, love, or nature, or disillusioned after losing faith or love: From outer bleakness, trial and testing, to inner bleakness and endurance.

This European mood also took hold in North America, both in the United States and Canada. The striking landscapes of late-nineteenth-century northern European painters, such as the Finn Akseli Gallen-Kallela, the Swede Eugene Jansson, and the Norwegian Harald Sohlberg, among others, influenced Canada's Group of Seven, who in the early twentieth century made visible the harsh beauty of their own northern landscape. At the same time, the Finnish composer Sibelius was writing music with strong affiliations to the lonely northern landscape. He, in turn, became a favourite of the Canadian Glenn Gould, whose "radio poem" *The Idea of North*, created a portrait-in-sound of the harshness and solitude of the North, one that concludes with a quotation from the triumphant finale of the Sibelius Fifth Symphony.

The Western world's obsession with the North also changed its focus on animal life, and on the image of "Man the Hunter." The wolf and the bear have always been mythically resonant, but after Darwin, these favourite animals of the North made a reappearance in mainstream literature, along with the notion of *Homo sapiens* the hunter. The northern circumpolar bear cults go back as far as 7000 B.C. The bear's hibernation may be one of the sources of the human idea of rebirth, and may have inspired the notion of the hero's journey to the underworld. Folk tales of human encounters with bears, notably the famous European tale of the Bearson, seem to have influenced the tale-tellers of *The Odyssey, Beowulf,* and *The Saga of Grettir.* The Viking berserkers or "bear-shirts" were warriors without armour, wrapped only in bearskins or wolf skins, an image that connects closely with the idea of the shape-shifter, and thus touches on much fantasy literature, from the fairy tales to Tolkien.

After Darwin, the wolf image became even more popular. Although he mistook the nature of its aggression, Jack London's *The Call of the Wild* of 1903 shows the fortunate reversion of the dog Buck to wild splendour as the leader of the wolf pack. Jack London constantly refers to the wolf, as do many writers of the time, in terms of awe and affiliation. This seems very appropriate, since, as anthropologists are now telling the story, the domestication of the wolf into dog by the naked ape coming out of Africa about 135,000 years ago may have been one of the key factors in giving those ancestors of *Homo sapiens* a chance to survive and prosper.

One of Stefansson's favourite "personae," one of the masks with which he faced the world, was that of the food gatherer. Stefansson's departure from the *Karluk*, one of the central moments of controversy in his career, was motivated, he tells us, by the desire to use his superior hunting skills to kill some game to feed his expedition party. And some of the most vivid moments in his *My Life with the Eskimo*, his most genial and characteristic book, are about hunting and tracking wolves, bear, and caribou.

On the surface, the world revealed by the northern imagination is often bleak and unyielding, inhuman and recalcitrant. It can suggest the hopelessness of the human struggle within nature, and generate a sense of pessimism, even of despair. On the other hand, the northern imagination finds beauty in a harsh environment, magic and mystery in the starkest landscape, and creates rituals and stories that evoke a spiritual dimension.

The northern imagination is an aurora at nightfall, a shaman's dance, howling wolves, cracking ice, trees and tundra, glaciers and moonlight, a penumbra of meaning around a half-forgotten cairn. The northern imagination suggests that nature is larger than humanity and doesn't exist merely to serve humanity, despite the improvised towns and the oil rigs, the pipelines and the paved roads.

The northern imagination reminds us of the power of simplification, of solitude, and at the same time of the importance of human contact and communication. It celebrates the acceptance of fate, even while challenging it, and points to survival itself as a value not to be despised.

In the future, assuming we begin to explore and settle the moon, Mars, and the other planets, visiting bleak landscapes and adapting to harsh environments, many aspects of the human experience shaped and revealed by the northern imagination will be revived in a new context and with new imagery and stories. In short, the northern imagination would seem to be a central and perennially relevant mode of human expression.

Yet there is a strong caveat to this analysis, a devastating footnote we must add to this survey of the phenomenon of "the North" as understood in past decades and centuries. For more recently we have encountered an aspect of change not anticipated by Stefansson or others in their efforts to convince us of the value of the North for the industrialized nations of the Temperate Zone. While scientists like René Dubos have taught us that "the spirit of place" is more than a figment of the literary imagination, that regional cultural development is often shaped by and linked to the enduring local environment, we are now more than ever aware of the global context. This may be summed up in a single phrase: Gaia is indivisible. Global warming is changing all regions of the Earth in ways that make assumptions of stability unrealistic. Stefansson's idea of a southern adaptation to the North becomes rather meaningless when the North changes its character completely. And that is precisely what is happening. According to a recent report of the Natural Resources Defense Council:

Between 1979 and 2008, the summer polar ice cap shrunk more than 20 percent. Average temperatures in the Arctic region are rising twice as fast as they are elsewhere in the world. Arctic ice is getting thinner, melting and rupturing. The polar ice cap as a whole is shrinking; the area of permanent ice cover is contracting at a rate of 9 percent each decade. If this trend continues, the Arctic could become ice-free by the end of the century. (NRDC Bulletin on Global Warming, 2008.)

In an issue of August 28, 2008, the *New York Times* featured a touching photograph of a polar bear, churning and thrashing, seemingly helpless, in the deep Arctic waters of the Chukchi Sea, northwest of Alaska, and just west of the Beaufort Sea region where most of Stefansson's work took place. The accompanying article describes how polar bears have been sighted, desperately trying to swim the four hundred miles of ocean they now must cross to reach their natural habitat of polar ice. The article also reported that melting of sea ice is not only changing the northern habitat in important respects but is also triggering another effect called "Arctic amplification." This is described as a feedback mechanism in which the warming that takes place up north is increased and the effects spill southward. In addition, the release of methane from the Arctic waters is increasing the Earth's dose of greenhouse gases even further. "We're moving beyond a point of no return," said the scientist heading a recent multinational scientific assessment of Arctic conditions.

Such changes, suggesting a radical transformation of traditional north–south polarities, and quite possibly offering a fundamental challenge to the development of Western civilization in the future, make Stefansson's predictions of northern development controlled by enlightened planners in the south seem rather naïve. The photograph of a polar bear, swept away on tides of melted ice, offers a near-caricature image, an ironical commentary on his idea of "the friendly Arctic." More and more, the idea of North propounded by Stefansson looks like an outdated proposition, a cultural image based on the passé notion of a permanent northern environment, and with every new discovery about the North, its grounding in reality seems more tenuous. If our developed civilization survives the radical changes it has seemingly precipitated in the Far North and around the world, it will look on the natural environment, and the northern environment in particular, in a new way, one not contemplated in any of Vilhjalmur Stefansson's bold and seemingly "visionary" projections.

And yet there is a final irony. If the Arctic warms and then the climate stabilizes, without disrupting the basic economic, social, and political systems of the circumpolar nations, then it is just possible that some of the developments foreseen by Stefansson will actually take place in Canada. The fantasy of cities ringing the pole, and resort beaches on the Beaufort Sea, may become a reality; a very different environment than the one imagined by Stefansson, but in some respects tantalizingly similar, and quite positive. Yet because Arctic warming is likely to do more damage than good, and given the indivisibility of the Earth's environment I referred to earlier, that rosy prospect seems more like science fiction than a potential reality. Was Stefansson right, or dead wrong, or right for reasons he never suspected? Only our grandchildren and great grandchildren will know the answer.

Stefansson the young explorer, casually poses on a dock.

Epilogue

Ice Follies: Dialogues in Limbo

Cast of Characters

Vilhjalmur Stefansson: Explorer

Harold Noice: Seaman, trail-mate of Stefansson, and sometime journalist

Fanny Hurst: Popular writer. For seventeen years Stefansson's lover

Dr. Rudolph Anderson: Zoologist, second-in-command during the Canadian Arctic Expedition

Admiral Robert E. Peary: Great American Arctic explorer, previously credited with being the first to reach the North Pole

Sir Ernest Shackleton: Great Anglo-Irish explorer whose planned Canadian expedition was blocked by the federal cabinet during the 1920s

Roald Amundsen: Great Norwegian explorer, first to reach the South Pole, and to transit the Northwest Passage

Dekoraluk: Stefansson's favourite dog

Scene

A wide, windless plain of ice and snow, like a blank white canvas awaiting a few master strokes to give it meaning. After a while, a man walks into the scene, a strong wiry man of medium height, dressed in skins and furs like an Inuk. A glimpse of his face shows well-defined features, Scandinavian or Russian, with a large nose, thick wide lips and penetrating blue eyes. Over his shoulder is slung an Austrian 6.5 mm Mannlicher-Schoenauer rifle, which he removes and sets carefully against a snowbank. He walks a few feet farther on, stops before another, larger snow bank and tests its strength by examining the depth of the imprints made by his soft deerskin boots. He nods with satisfaction, then pulls from inside his jacket a twenty-inch machete and begins to cut the snow into domino-shaped blocks.

After a while, another figure enters the scene from the right, a man somewhat younger than the first, also dressed in the clothing of

an Arctic explorer of long ago. This man, sturdy and muscular, wears an uneasy, slightly guilty look. He glances at the man building the snow house, hesitates, and then, seeming to take courage, speaks.

Noice: Hello, Stef! What's the matter, don't you want to talk to me? I know you saw me arrive — nobody can creep up on you. I remember that famous polar bear incident you told me about ...

Stefansson: *(looking up, with a sharp smile)* Hello, Noice. You mean that bear that growled a little just before he was about to spring and kill me? Yes, I was lucky that time. I had my back to him. Unfortunately, some of my enemies didn't serve notice like that.

Noice: You're not suggesting you would have shot them? I know that's not true. You're just not a violent man. Did I ever tell you what I heard about you before we met? It was from the whalers I hung out with at Herschel Island. They said you were a college professor who carries books in his sledge when a sensible man carries grub. They asked me why I wanted to work with a man who doesn't drink, doesn't smoke, and doesn't dance. You didn't care about sports, they told me, never opened your trap to hum a tune or make a dirty joke, You didn't play cards or make bets. The worst they ever heard you swear was a good loud "Gee whiz!" When I met you I wasn't impressed — that soft voice of yours, those lily-white hands. When that ship of ours, the *Bear*, got stuck in the ice, you read a novel. You didn't bat an eye. I was sure we would starve to death. Then you got going with that rifle of yours ...

Stefansson: *(laughs)* You were all right, Noice. You were willing to try out my methods. I knew I could feed everyone with my Mannlicher. I was always getting resistance from the fellows on the trail. You and Karsten Anderson were different. That's why you shocked me so much later.

Noice: *(wearing a hangdog look)* You mean about the Wrangel Island stories? You know how badly I felt about that. I got carried away. Like I told you in New York, they garbled some of my interviews. I — I guess it went to my head — all that attention. I always wanted to be something more than a sailor. I had some education. I felt I could tell a story. And what I saw on Wrangel really shocked me. I had to talk about it. And the newspapers asked for more. Always more.

Stefansson: Didn't it ever occur to you how it would reflect on me? Why do you think I sent you? Do you think I wanted them to die? Maurer was a good man, and Knight, and that young student, Crawford. They were all decent fellows. Do you think it was easy for me? Have you ever tried to get quick action from a government when you most need it? That Mackenzie King, I called him "my old Harvard friend," but he didn't come through for me. Neither did Billy Mitchell and the Americans. They should have put an air base on Wrangel. Just think how it would have served them during the Cold War … Governments — I got fed up with them all! And those damned McCarthy-ites! Or should I say, Gee whiz? Maybe I should have stuck to building ice houses — that's something I'm good at. *(crouches, continues cutting ice blocks with his machete)*

Noice: I understand, Stef. Believe me, I sympathize. But you were never on Wrangel. It's nothing like a paradise. It's a bleak, goddamn hell-hole. Or was. Only the Soviets, with all their Siberian experience, could make anything of it. And when I arrived with the rescue ship it was a mess. Three men missing on their way to Siberia to get help. Hunger drove them out on the ice, and they must have fallen through. Nobody's ever found them. And that Eskimo woman with the funny name — Ada Blackjack — doing everything for everybody, and even she had scurvy. She seemed half crazy, when I got there, pawing over the body of Knight. He was a guy you respected, I know. So did I.

Stefansson: *(stands up but does not look at Noice)* I don't need to hear any more, Harold.

Noice: I still have nightmares about it.

Stefansson: I told them to get an umiak. You can't get many walrus without a skin boat. I told them to take along a couple of Eskimo families. They needed more experienced hunters. And the relief ship didn't get through in the first year. The key was that Knight got sick. He didn't follow my ideas, or he wouldn't have gotten scurvy. He could have made it across the ice to Siberia, but the scurvy got him. So the others tried the trek, and died. They made mistakes. None of them should have died.

Noice: *(starting to shiver)* It's getting cold. Besides, I see someone else is coming to talk to you. My time is up, I guess. You only get so much time in this world. I'm sorry it happened the way it did. I wish we could do it all over.

Stefansson: So do I. Unfortunately, it doesn't work that way.

(Noice pulls his coat around him and walks off toward the left. Stefansson stops his work for a moment and watches him. Neither man speaks. Noice disappears. Stefansson stands, head bowed, eyes closed, the machete gripped tightly in his right hand. When he hears a woman's voice calling his name from close by he looks up.)

Fanny: Stef!

(Stefansson puts down the machete. A tall dark-haired woman, elegantly slim, and dressed in a mink coat, with a flowing cashmere scarf and matching fur hat, approaches from the left. She stops only a few feet away from the explorer. For a few long seconds, they gaze at each other in silence.)

Fanny: You're looking well, my love. Time has treated you kindly.

Stefansson: And you're as beautiful and as chic as ever. I was hoping you would come by. Hoping and fearing.

Fanny: Fearing? Afraid you might fall in love with me all over again? That would be too good to be true, but as we know, love doesn't work that way. I've never seen you like this, in your trail outfit, except in photographs. You look just as heroic as I always imagined. Of course you looked heroic in Abercrombie and Fitch gear too. And dressed by Brooks Brothers. Or naked, especially naked. Why are you cutting those ice blocks?

Stefansson: I'm building a snow house, Fanny. I told you about those.

Fanny: *(laughs nervously)* Oh yes — you told me the temperature was so warm inside that at night you could strip naked, crawl under the covers, and stay warm. You could be comfortable and your clothes could dry. I wish we could have made love in a snow house.

Stefansson: *(frowning)* It was much better in your Manhattan apartment.

Fanny: *(hesitates)* Yes … I guess the snow house was reserved for that Eskimo wife of yours. Oh, don't look so defensive! I was never jealous. Of that fat thing? Oh I know, don't look so ferocious — Eskimos aren't really fat. They just wear heavy clothing. Or so you say. But damn it, she was fat! For a while after you mentioned her — did you really call her Fanny? — the idea excited me. It made me want you more. You and your Eskimo woman. Why didn't you bring her to New York? That would have been a sensation. I could have written a piece in *Harper's* defending her rights as a woman. NO MORE TRIBALISM! FREE WOMEN FROM PREDATORY WHITE EXPLORERS! Or was she a blond Eskimo? I never thought they looked very blond, by the way. Good for them!

Stefansson: Don't be cruel, Fanny. You know I don't want to talk about that. I'm sorry I ever told you. I remember it was one of those nights when lovers tell each other everything. I'm sorry we had to part. We had a long time together.

Fanny: Seventeen years! Longer than many marriages — longer at least than those of most of my friends. You were the best thing that ever happened to me, Stef.

Stefansson: I don't believe it; your talent was the best thing that ever happened to you. And your personality. Everybody loved your panache.

Fanny: It wasn't enough to hold you.

Stefansson: It was you that really broke things up, Fanny. I would have married you, but you were afraid.

Fanny: I was never afraid of marriage — with you. I was afraid of being poor. And besides, you fell in love with a younger woman. You wouldn't have married me anyway. The same old story. If I had been a gorgeous Ziegfield girl and had thrown you over for someone richer, then you would have pined for me. That would have been the smart thing for me to do. And why shouldn't I have done the smart thing? I was part of the Smart Set. But I did the dumb thing. I loved you more than you loved me.

Stefansson: On your own terms … I don't want to hurt you, Fanny, but yes, I did fall in love. Our love was a beautiful old thing by then, still shiny but a bit worn. With you I would have been giving hostages to fortune. With her, marriage was insurance against death.

Fanny: Insurance against death?

Stefansson: Don't you understand? The idea of death disturbed me; not when I was on the trail, trying hard to survive, but later, after the exploring stopped. I began to feel old, not up to the life of adventure. Time became my enemy. The best antidote to that is erotic fixation; at least I've found it so. But you and I had cooled down too much by then. Death was getting a grip on my vital soul.

Fanny: I can't believe that, Stef. I always celebrated your lovemaking. I even bragged about it to my girlfriends — as a fan, as a partner. I can't believe what you say. We never cooled off. You chose her because she was young, and very pretty, and intelligent enough. But that's all done with now.

(They are silent for a few long seconds. Fanny bows her head, turns, and starts walking away, left, then stops suddenly.)

Fanny: It was the money! I couldn't get Jacques to let me go. Our marital roles had been reversed long before; he was the housekeeper, I was the moneymaker, the star. He was a good pianist but he was no Rubenstein. If I'd divorced him, he would have milked me for everything he could get, and I couldn't imagine living on your income, or letting you manage mine. I bitterly regret it now.

(She walks on, not looking back. Stefansson watches her closely, and so fixedly that he doesn't notice a man who enters the scene from the right. Strong and compact, he is about Stefansson's height, but broader. Dressed in old-fashioned Inuit clothing, he has a long, handsome face, with a high forehead, deep-set penetrating eyes,

and a small, carefully clipped dark moustache. When he sees Stefansson he frowns and starts to walk on.)

Stefansson: Anderson! What are you doing here? I see you have no greeting for an old trail-mate.

Anderson: A trail-mate? That's a laugh. I was never your "trail-mate" and you know it. I didn't enjoy your company — not ever — any more than I liked your methods.

Stefansson: I didn't want to contend with you, Rudolph. I never understood why you hated me so much. You would never have gotten to the Arctic without me. Perhaps that's the problem. I invited you there, and gave you the means to do your work. You never let me do mine.

Anderson: What work? Shooting polar bears? Playing the hero? Rambling around a few obscure islands? You never did any respectable intellectual work. You accomplished nothing in that line. At self-promotion, though, you were a master.

Stefansson: As bitter as ever, I see. You know, I never understood why you were so fixated on me. Your obsession was obvious. Everything I did was suspect. You went out of your way to undermine me. And your wife, the egregious Mae Belle — she soon picked up on it. She was even more obsessed than you. You never should have married her, Rudolph. I'm beginning to see that.

Anderson: *(takes a step toward him, threatening)* How dare you? If you didn't have that machete and that rifle …

Stefansson: I didn't have any hatred for you, Rudolph, not a trace. In fact, whenever you left me alone, I didn't think about you very much. I had more important things to do. You should have cooperated with me. Too bad you didn't trust me. I could have helped you.

Anderson: *(exasperated, hardly able to speak)* You ...!

Stefansson: Your venom amazed me. I almost laughed. That letter to the Explorers Club ... people have been committed to asylums for less. To call me a "bounder"... a "cosmopolitan superman above country." To say that I "sponged my education." That I was "a rooster fighting for my dunghill." To refer to my "yellowness" that was "congenital." By heaven, you even called me "an international Socialist." Oh, no, anything but that! Senator McCarthy could have learned something from you.

Anderson: Damn you! You were "loosely educated," as Churchill said of Hitler. Oh, I know you were a great reader, but you had no real scientific training. And your loyalties were always suspect. When I went to Canada I found a home. I believed in my new country and in Western democracy. As soon as you couldn't run the show in Ottawa, you deserted. I saw through the Soviet scam; you never did. You concealed your leftist leanings in your autobiography. You always arranged everything so as to make yourself look good. You couldn't organize, you couldn't take responsibility, you had no patience for real work. You were a failure, only saved by your talents as your own press agent. I had more to give than you ever did, but I hated self-promotion.

Stefansson: It's true that you were a terrible speaker, Rudolph. And a diffident, surly fellow at times. And, to put it mildly, not a very generous colleague. But I tried to keep things in perspective. There are always a lot of people like you in this world; they hang out in universities, mostly. Or in government. They skulk around, complaining about everything, attacking people who stick their necks out and do anything. They seethe and burn, but they can never manage a creative act; they can only criticize the actions of others. Unlike them, unlike you, I took chances. I stuck my neck out. I had a vision. I made mistakes, yes, but I got some things done. Canada may have rejected me, but I gave them plenty to think about. I made them see their own North, which they had pretended didn't exist. It's something they'll have to deal with. My ideas are as provocative as ever. I may have been wrong in some cases but I knew what was important.

Anderson: The future will discredit you and everything you stood for. Scholars are already picking holes in your "heroism" and your "vision of the North." And men died because you jabbered so much about "the friendly Arctic."

Stefansson: You had your chance, Rudolph. Unfortunately, you wasted too much energy in jealousy. Isn't that one of the cardinal sins? The sad fact is that, in the future, when anybody thinks of you, it will only be in connection with me. That's the hell you prepared for yourself by your fixation. You've become my Iago, my dark shadow twin. I feel sorry for you, Rudolph.

Anderson: *(still seething)* Feel sorry for yourself; it's you that has to face the Judgment.

Stefansson: *(looks baffled)* The Judgment?

(Anderson shakes his head and storms away, exiting left, without looking back. Stefansson watches him thoughtfully for a while then shrugs his shoulders and picks up his machete. As he begins to chop yet another ice block, the light changes. The sky darkens, the horizon closes in. Snow swirls about him. The wind howls. As Stefansson turns from his work, a heavy fog starts to envelope him. He peers into the fog, and suddenly, toward the right, a few huddled shapes appear. One by one they struggle out of the fog and become visible.)

Stefansson: *(to himself)* Now what's happening? This is a strange place.

(The first shape stands before Stefansson. It is a broad, burly mountain of a man, dressed in heavy skins, like an Inuk, with high white boots and a bulky jacket, his weather-beaten face mostly hidden by a large walrus moustache.)

Stefansson: Peary!

Peary: *(for a moment looking baffled)* Who…? Ah, yes, Stefansson. I remember. I should have known you'd be here … But I must go on. It seems that even here one must defend oneself. They want to talk to me about my sighting of "Crocker Land," which of course never existed. Then there's Matt Henson. Damn you, man, did you fellows have to give that Negro a medal? He was my manservant, for God's sake! All right, I couldn't have reached the pole without him. And, by God, I did reach it! As for my sightings and compass readings — it's true they might have been slightly off.

Stefansson: What does standing right at the pole matter? When people asked me if I ever reached the pole, I told them: "I'm not a tourist!" You were a hero, a great explorer.

Peary: *(brushing snow from his coat)* And after my death, you told them I had a ghost writer! To suggest that I was depressed and not responsible for some of my actions when I wrote about my journeys. You let me down, Stefansson, and after I got off my deathbed to shower you with praise. It was ungrateful of you. But I must go now.

Stefansson: Wait, Peary! I can explain …

Peary: It's all behind us. It doesn't really matter.

(Peary tramps away, exiting left, without looking back. Stefansson, frustrated, kicks at the snow. Another figure emerges from the mist, a stocky, dark-haired, handsome man, whose intelligent eyes belie his look of an old-fashioned Irish boxer.)

Stefansson: Shackleton, you here too?

Shackleton: *(stops in his tracks)* I know I should know you, but I can't quite remember.

Stefansson: Stefansson is the name.

Shackleton: *(a vague smile)* Ah, yes, the Canadian bloke with the Icelandic face and name. I remember all right. I put you in touch with my press agent, gave you some very good advice about promoting yourself in America, and then, when the chips were down, you blocked me out of Canada.

Stefansson: *(shocked)* Please understand, I had nothing to do with that! It was the damned Canadian politicians! Mackenzie King was afraid to commit himself to anything doubtful or risky. Not a man to take chances. As for that Ellesmere Island Expedition — they couldn't choose between us, but were happier with neither. Inaction was their specialty. I left Canada for the States in disgust some years later. By then my exploring days were over.

Shackleton: I'm sure you must have done well in the States. We all admired you — or rather your exploits. And you're here, aren't you? That tells me something. Only the great and successful have to face what we do. You're one of us, Stefansson. I suppose you'll be joining us soon — over there. This is a strange world. I hope there's still something to discover.

Stefansson: *(thoughtful, hardly looking at him)* There must be …

(Shackleton nods, trudges slowly away, and disappears. The last of the half-hidden figures emerges from the mist. Unlike the others, he is travelling on skis. Dressed in Inuit furs, a tall, strong man with a high forehead, compelling Nordic features, and a penetrating gaze, he seems not to notice Stefansson, and is nearly past and gone, before the latter calls out to him.)

Stefansson: Amundsen!

Amundsen: *(the Norwegian stops, turns to Stefansson and looks at him for a long time)* Who are you? I am in a hurry, and have no time for chatter.

Stefansson: *(with a wry smile)* I doubt if you will be happy to see me again. We did meet once, long ago, on Herschel Island. It was my first expedition. You were not yet famous. Much later, after much Arctic travel, I published a few books, and you didn't like them. You said that the first of my discoveries, the so-called "Blond Eskimos," was "the most palpable nonsense that has ever come from the North." And my second idea, that a skilled man could live off the land, even in the Arctic, you called "harmful and dangerous nonsense." I'm Stefansson.

Amundsen: *(erupts in anger, waving one of his ski poles)* If you're Stefansson, you're a bloody fool! A curse on your "friendly Arctic!" I died up there, in the snow and ice! Yes, I survived the crash, but I wish I hadn't. It took that much longer to die. The North was full of deceptions; the Antarctic was simple by comparison. You have a lot to answer for, Stefansson!

Stefansson: I proved what I proved. One can adapt to almost any environment. I only followed your methods, although I never would sanction eating my dogs. I don't think you ever read my book. One of your disciples probably told you that, buried in my account, there was a slighting reference to your transit of the Northwest Passage. It wasn't meant to be a put-down. I was just expressing the usual qualification.

Amundsen: You should have qualified some of your own nonsense. If you'll excuse me, sir, I must go on now. I must say, meeting you has not improved my humour.

Stefansson: *(shrugging his shoulders, then speaking sotto voce)* I can't believe your humour was very good to begin with.

(Amundsen disappears in the whirling snow. Stefansson stands, swinging his machete from side to side, looking uncertain and rather sad.)

Stefansson: *(muttering to himself)* I suppose I must go on.

(Another shape springs out of the whirling snow. It bounds in from the right, and makes straight for Stefansson — a fine husky dog, looking sleek and well fed, full of life energy and affection. Stefansson gives a cry of recognition, tosses away his machete, and embraces the animal.)

Stefansson: Dekoraluk! My old companion! My best dog!

(The two romp for a while in the snow. Then Stefansson, smiling, claps his hands together, and starts off left, in the direction the others took. Dekoraluk runs along beside him.)

Stefansson: This is much better. A true trail-mate at last! Ah, yes, my old friend. You were always friendly, always faithful. We often worked hard together — you much harder than I, and more cheerfully. We've often been hungry together, haven't we? You even more than I! But you never shirked, you never gave up. You were honourable and magnanimous — as a dog can sometimes be! Refusing to steal meat from our cache, refusing to fight, although you were the best fighter of all our dogs. Your death took my courage away — for a little while. But now you're back, and we can take the trail again, together. Come along, old boy, let's go on, you and I — into the great white world, into the unknown.

(Man and dog disappear into the swirling snow.)

THE END

Chronology of Vilhjalmur Stefansson (1879-1962)

Compiled by Tom Henighan

Vilhjalmur Stefansson	*Canada and the World*
	1847 Captain Sir John Franklin dies while leading a British expedition to the Canadian North, and all his men subsequently perish after abandoning ships trapped in the ice near King William Island.
	1853–54 Dr. John Rae discovers relics of the Franklin Expedition in possession of the Inuit. He is publicly mocked when returning to Britain for trusting the word of "savages."
	1857 Queen Victoria names Ottawa as the new capital of Canada on December 31.

Vilhjalmur Stefansson

Canada and the World

1860–61
Isaac Hayes leads an American expedition in search of the legendary Open Polar Sea, which is proved to be a myth.

1860–62
American Charles Francis Hall makes his first journey to the Arctic in a search for any survivors from the Franklin Expedition. He discovers relics from the Frobisher journey of 1576–77.

1864–69
Charles Hall makes his second journey to the Arctic. He lives and travels with the Inuit by sledge across Rae Isthmus to King William Island where he finds artifacts from the Franklin Expedition.

1867
On July 1, the Dominion of Canada, uniting Ontario, Quebec, New Brunswick, and Nova Scotia, comes into existence, with John A. Macdonald as first prime minister.

1870
On May 15, Manitoba becomes Canada's fifth province.

1871
The first census of the Dominion of Canada lists the population as 3,689,257.

Vilhjalmur Stefansson	*Canada and the World*
	British Columbia enters Confederation on July 20 as the nation's sixth province.
	1873 Prince Edward Island enters Confederation on July 1.
	1876 The first telephone call between separate buildings is made by inventor Alexander Graham Bell, in Mount Pleasant, Ontario, to his uncle, David Bell, in Brantford, Ontario on August 3.
1879 Vilhjalmur Stefansson born on November 3, in Árnes, Manitoba, north of Winnipeg.	**1879** Sir Sandford Fleming presents a paper on February 8 to the Royal Canadian Institute proposing that the world be divided into twenty-four time zones.
1881 The Stefansson family moves to a small Icelandic settlement in the Red River Valley in Pembina County, North Dakota.	**1885** Rail director Donald Smith drives the ceremonial last spike home for the Canadian Pacific Railway, linking Montreal to Port Moody, British Columbia on November 7.
	1885 On November 16, Métis leader Louis Riel is hanged for high treason as a result of the Northwest Rebellion.

Vilhjalmur Stefansson	*Canada and the World*

1886
Robert Peary attempts to cross Greenland but fails.

1888
Fridtjof Nansen successfully completes the first Greenland crossing.

1891–92
Peary's second expedition to Greenland.

1893–95
Peary's third expedition to Greenland.

A new farthest north record is established when Fridtjof Nansen and Otto Sverdrup, in the *Fram*, drift across the Arctic Ocean.

1897
Salomon Andrée, and two companions aboard the balloon *Eagle*, attempts to reach the North Pole. They abandon the balloon after a short flight. They disappear and their preserved bodies are found on White Island, off the northeast coast of Spitzbergen in 1930.

1898–1902
Peary's third expedition to the Arctic. His plans to reach the North Pole end in failure.

Vilhjalmur Stefansson

1899
Stefansson enters the University of North Dakota at Grand Forks as a freshman, after taking various make-up courses to compensate for his lack of any high school preparation.

1903
Stefansson graduates from the University of Iowa, having transferred there after being suspended by the University of North Dakota for being a "trouble maker."

He begins his study at Peabody Museum and in the Anthropology Department of Harvard University.

1904
He makes a summer trip to Iceland with a grant to

Canada and the World

1899
On October 30, more than 1,000 Canadian soldiers set sail from Quebec to South Africa and the Boer War.

1900
Liberal Wilfrid Laurier becomes prime minister on November 7 after defeating Charles Tupper's Conservatives.

1901–02
The first Ziegler Expedition, led by Evelyn Baldwin. The expedition attempts to reach the North Pole via Norway but ends in failure, one of many failures by various polar expeditions during this era.

1903
On October 19, Canadian representatives on the Alaska Boundary Commission refuse to sign the commission's decision setting the boundary between Alaska and Canada, saying virtually all American positions had been accepted.

1903–05
Roald Amundsen successfully completes the first navigation of the Northwest Passage aboard *Gjoa*.

Vilhjalmur Stefansson	*Canada and the World*

Vilhjalmur Stefansson

investigate the relationship between tooth decay and cereal eating in a selected Icelandic population.

1905
Second visit to Iceland. Digs up graves and collects skulls for anthropological examination.

In Boston, he meets Orpha Cecil Smith, whom he seems destined to marry.

1906–07
Joins the Anglo-American Expedition for his first fieldwork and travel in the North.

1908–12
With Rudolph Anderson, undertakes a second northern expedition. Ethnographic observations in diaries during these expeditions and the works published later reveal Stefansson as a pioneer observer of the North.

1908
Stefansson first meets and hires Pannigabluk, an Inuit seamstress, who will become his sexual companion in the North.

Canada and the World

1905–06
Peary's fourth attempt to reach the North Pole. His attempt only succeeds in establishing a new farthest north.

1905
Acts proclaiming Alberta and Saskatchewan as Canada's newest provinces receive royal assent on July 20.

1908–09
Peary's fifth and final attempt to reach the North Pole. Peary's vessel, *Roosevelt*, commanded by Captain Robert A. Bartlett, sets a record latitude for a ship under its own steam (82° 30' N). In March 1909, Peary claims to have reached the pole, though this has been since questioned.

Frederick Cook claims to have reached the Pole in April.

1909
John Alexander Douglas McCurdy makes the first airplane flight in the British Empire on

Vilhjalmur Stefansson

1910
Alex, the son of Pannigabluk and Stefansson is born on March 10.

Stefansson visits the Copper Inuit on Victoria Island. Reports finding some individuals with white racial features, which results in almost instant fame for the explorer.

1911
Cecil Orpha Smith marries her childhood sweetheart.

1913–18:
The Canadian Arctic Expedition. The *Karluk* disaster takes place, and there is a falling out between Stefansson and Rudolph Anderson. Several previously unknown islands visited and claimed for Canada, and much scientific research accomplished, in particular by Diamond Jenness and the Southern Party.

1913
My Life with the Eskimo published.

Canada and the World

February 23, travelling about ten metres above the ground for almost a kilometre at Braddeck, Nova Scotia.

1912
Ottawa divests itself of responsibility for vast tracks of northern land, granting boundary extensions to Manitoba, Ontario, and Quebec on May 14.

1913
Prime Minister Sir Robert Borden approves federal government backing of Canadian Arctic Expedition, led by Vilhjalmur Stefansson, which will last for five years and claim new Arctic territory for Canada.

1914
Following Germany's invasion of Belgium, Britain declares war on Germany on August 4. Canada, as part of the British Empire, is engaged in the war as well.

1917
On April 9, the Canadian Corps attacks German positions

Vilhjalmur Stefansson	*Canada and the World*
	on Vimy Ridge in France, a key piece of land held by the Germans since 1914. Six days later, fighting ends with the Canadians victorious despite the loss of 3,600 troops.
	Mont Blanc, a French munitions ship, explodes in Halifax Harbour on December 6, killing more than 1,000 people and destroying some 6,000 homes.

1918

Stefansson settles in New York, at the Harvard Club, and finally in Greenwich Village.

Stefansson begins what are to be many years of cross-continent lecturing on the subject of the North.

1918

On May 24, Canadian women win the right to vote in federal elections.

The First World War ends on November 11; Canada has lost 60,000 troops.

1919

On May 15, a general strike begins in Winnipeg in support of striking workers in building and metal trades. It ends six weeks later, after two deaths in skirmishes.

1920s

New York joins London and Paris as one of the world's greatest cultural capitals. Greenwich Village becomes the centre of much new art, including jazz, modern painting,

Vilhjalmur Stefansson

Canada and the World

theatre, fiction, and fashion, while radical political thinking flourishes and the legendary "roaring twenties" initiate nearly a decade of uninhibited freedom from conventional American mores.

1920
The Royal North West Mounted Police and Dominion Police merge on February 1 to form the Royal Canadian Mounted Police.

Group of Seven painters begin to depict the Canadian wilderness and the North in a newly expressive style.

1921
The Friendly Arctic is published, summarizing the 1913–18 Expedition. Despite some negative reactions, especially in Canada, Stefansson is honoured by the most important geographic societies in North America and Europe, and officially cited by the Canadian government.

1920–21
Canadian women gain the right to serve in the House of Commons and Agnes Macphail becomes the first woman elected.

1921
The Lunenburg fishing schooner *Bluenose* defeats the American vessel *Elsie* to win the international schooner championship on October 24.

1921–23
The Wrangel Island settlement ends in tragedy and scandal.

1922
Stefansson meets Fanny Hurst in Ravello, Amalfi, Italy. Their seventeen-year relationship begins.

1921–22
Between December 6 and January 3, The Royal Mint produces Canada's first five-cent pieces, made mostly of nickel.

Vilhjalmur Stefansson

1924
Stefansson travels to the
Australian deserts.

Canada and the World

1923
On October 25, Frederick
Banting and J.J.R. Macleod are
first Canadians to win a Nobel
Prize for their work that led to
discovery of insulin.

1925
Amundsen's twin seaplane flight
from Spitzbergen lands on an
ice floe, 136 nautical miles short
of the North Pole. One of the
planes manages to take off again
but is aborted at sea near North
Cape, Spitzbergen, and the crew
rescued by a sailing ship.

1926
On May 8, Floyd Bennett
and Richard E. Byrd claim to
have flown to the North Pole.
However, despite ticker tape
parades and the Congressional
Medal of Honor for Byrd, he later
admits privately that they did not
actually reach the pole.

The Amundsen–Ellsworth North
Polar dirigible flight reaches the
North Pole in the early hours of
May 12, Ellsworth's birthday.

On November 19, The Commonwealth adopts the Balfour Report,
specifying that dominions such
as Canada are autonomous from
and equal to Britain

Vilhjalmur Stefansson

1927
Stefansson helps organize and becomes a subject in controlled experiments on all-meat diet in New York.

Canada and the World

1927
The British dominion of Newfoundland wins a twenty-five-year boundary dispute with Canada. Labrador, which had been claimed by Quebec, is awarded to Newfoundland on March 2.

1928
Umberto Nobile leads an all-Italian expedition to the North Pole aboard the dirigible *Italia*, which ends in disaster. Roald Amundsen and four companions are killed in a plane crash during a rescue attempt.

George Hubert Wilkins, Australian-born former companion of Stefansson, flies with Carl Eielson across the Polar Sea from Point Barrow, Alaska to Spitzbergen in 21 1/2 hours.

On April 24, the Supreme Court rules that women are not persons, and therefore are not eligible to sit in Senate. The government later amends the British North America Act to allow women to enter Senate.

1929
On October 29, the New York Stock Market crashes.

Vilhjalmur Stefansson

Canada and the World

1930
After negotiations with Ottawa, Alberta gains control of its natural resources on October 1. Saskatchewan and Manitoba also receive the same power this year.

1931
On July 6, federal officials and the Red Cross announce plans to aid victims of a drought that has gripped the Prairies for more than a year.

The Statute of Westminster, giving dominions of the Commonwealth full legal freedom, is passed by British Parliament on December 11. At Canada's request, Britain retains power to amend the British North America Act.

1932
Legislation brings the Canadian Radio Broadcasting Commission into existence on May 24.

Canada and the United States agree to develop the St. Lawrence River into a seaway capable of taking ships into the Great Lakes.

Hubert Wilkins attempts without success to cross the Arctic Ocean by submarine.

Vilhjalmur Stefansson

Canada and the World

1933
Franklin Roosevelt sworn in on
March 4 as the thirty-second
president of the United States.

1934
Parliament passes the Bank of
Canada Act on July 3, creating a
central bank.

1939
Canada declares war on Nazi
Germany on September 10.

1940
Pannigabluk dies of tuberculosis
in the Arctic.

1941
Stefansson marries Evelyn
Schwartz-Baird, and they move
to a farm in Vermont.

1941
On June 27, the federal
government agrees to allow
women to enlist in the army.

Canada declares war on Japan
on December 7 after its attack on
Pearl Harbor.

1942
All Japanese on Canada's West
Coast are to be moved inland
to camps, the government
announces on February 26.

Canadians voting on April
27 in a plebiscite support
conscription, but the vote badly
divides the country: 70 percent
of Quebecers reject it.

On May 11, a German U-boat in
the St. Lawrence River torpedoes

Vilhjalmur Stefansson

Canada and the World

two freighters, the first time
the war has come to Canadian
territory.

Canadian troops sustain major
losses on August 19 in an ill-
fated raid on the French port of
Dieppe. Nearly 1,000 Canadians
die and another 1,800 are taken
prisoner.

1944
In the largest amphibious
operation in history, Allied
troops storm the beaches
at Normandy on June 6
—Canadians take Juno Beach.

T.C. (Tommy) Douglas leads the
CCF to power in Saskatchewan
on June 15, becoming Canada's
first socialist premier.

1945
Victory comes for the Allies
in Europe as the Germans
surrender on May 7. News of V-E
Day touches off wild celebrations
in Canada.

The United Nations is established
on June 26.

First nuclear bomb is dropped on
Hiroshima, Japan, on August 6.

The Japanese emperor announces

Vilhjalmur Stefansson	*Canada and the World*
	Japan's surrender on August 15, ending the Second World War.
1946 Stefansson is contracted by the United States Navy to create a twenty-volume Arctic encyclopedia.	**1946** The Canadian Citizenship Act is passed on May 14, meaning a Canadian citizen is no longer classified as a British subject first. The government introduces Canada Savings Bonds on October 14. On October 17, Winston Churchill proclaims "an iron curtain has swept across the continent [of Europe]." The Cold War begins.
1948 The Navy cancels funding for the encyclopedia.	**1947** Drilling begins on February 13 at Leduc No. 1, a huge oil find in north-central Alberta. **1949** Newfoundland officially enters Confederation on March 31. On April 4, NATO is established as bulwark against Soviet expansion.
1950 The public prosecutor of New Hampshire interrogates Stefansson about possible Communist connections.	**1950** Canada enters the war between North and South Korea.

Vilhjalmur Stefansson

Canada and the World

1951
Parliament passes a motion on May 7, seeking a constitutional amendment that would create pensions for all Canadians over seventy years of age.

1952
Canada's first television station, CBFT Montreal, begins broadcasting on September 6.

1953
Stefansson joins Dartmouth College faculty. His extensive collection of Arctic material is passed on to Dartmouth's Baker Library. An Arctic Institute is established at the college.

1956
On June 6, a pipeline bill authorizing the creation of a western section of pipeline to transport natural gas to Ontario from Alberta passes second reading in the Senate. The bill causes an uproar after the Liberal government invokes closure — a time limit on debate — for the first time in history.

1957
John Diefenbaker's successful election campaign includes promise of northern development.

1958
The world's first nuclear powered submarine, USS *Nautilis* becomes the first submarine to reach the North Pole.

On June 26, Queen Elizabeth,

Vilhjalmur Stefansson

Canada and the World

prime minister John
Diefenbaker, and U.S. president
Dwight Eisenhower officially
open the St. Lawrence Seaway.

1960
First transit of the Northwest Passage by USS submarine *Sea Dragon*.

On August 10, the Bill of
Rights, specifying the rights of
Canadians, becomes law.

1962
First totally submerged transit of
the Northwest Passage (eastward)
by USS submarine *Skate*.

1962
Vilhjalmur Stefansson dies of
complications of a stroke on
August 26.

On January 19, the government
announces a new immigration
policy intended to remove any
racial discrimination from the
system.

1964
Stefansson's autobiography
Discovery is published.

1966
Alex Stefansson dies in Inuvik.

Saskatchewan's Medical Care
Insurance Act takes effect
on July 1, creating Canada's
first comprehensive public
healthcare program.

On October 22 the Soviet Union
pulls its missiles out of Cuba.

1967–68
Glenn Gould radio documentary
The Idea of North is created and
broadcast on the CBC.

Sources Consulted

Stefansson wrote innumerable articles, and to judge by only a sampling, most of them are readable and interesting in various ways. For a list of these and other material, see the biography by William R. Hunt, included below. Stefansson, in collaboration with others, wrote various books for younger readers, of which I have listed only one. To guide the reader who wants to dig deeper, I have offered brief descriptions and evaluations of the major books about Stefansson.

Books by Vilhjalmur Stefansson

(The books that are likely to most interest the contemporary reader, or are most central to Stefansson's life and work, I marked with an asterisk.)

*My Life with the Eskimo. New York: Macmillan, 1913.

Anthropological Papers. New York: American Museum of Natural History, 1914.

The Friendly Arctic. New York: Macmillan, 1921. Reprinted, with some new material, 1944.

Hunters of the Great North. New York: Harcourt Brace, 1922.

The Northward Course of Empire. New York: Harcourt Brace, 1922.

Kak, the Copper Eskimo (with Violet Irwin). New York: Macmillan, 1924.

The Adventure of Wrangel Island. New York: Macmillan, 1925.

My Life with the Eskimo (abridged). New York: Macmillan, 1927.

The Standardization of Error. New York: W.W. Norton, 1927.

"Adventures in Diet" (article) *Harper's* Monthly Magazine, November 1935.

Adventures in Error. New York: Robert M. McBride, 1936.

The Three Voyages of Martin Frobisher (in collaboration with Eloise McCaskill). London: Argonaut Press, 1938.

Unsolved Mysteries of the Arctic. New York: Macmillan, 1939.

Iceland: The First American Republic. New York: Doubleday, Doran, 1939.

The Problem of Meighen Island. New York: Privately printed for Joseph Robinson, 1939.

Ultima Thule. New York: Macmillan, 1940.

Greenland. New York: Doubleday, Doran, 1942.

Arctic Manual. New York: Macmillan, 1944.

Compass of the World (with Hans W. Weigert). New York: Macmillan, 1944.

Not by Bread Alone, 1946. (Enlarged edition issued under the title *The Fat of the Land.*) New York: Macmillan, 1956.

*Great Adventures and Explorations (in collaboration with Olive Rathbun Wilcox). New York: Dial Press, 1947.

New Compass of the World (with Hans W. Weigert and Richard Edes Harrison). New York: Macmillan, 1949.

*Discovery: The Autobiography of Vilhjalmur Stefansson. New York: McGraw-Hill, 1964.

Books about Vilhjalmur Stefansson

(listed in chronological order)

Earl Parker Hanson. *Stefansson, Prophet of the North*. New York: Harper Brothers, 1941.

Hanson was a Stefansson enthusiast, and a friend of the explorer. His book, as he explains, is largely based on material collected by D.M. LeBourdais for a long biography of "Stef" that was never published (see the next entry). Hanson plays Lowell Thomas to Stefansson's T.E. Lawrence, but, as in the case of Thomas, the result is very readable, if tending toward hagiography.

D.M. LeBourdais. *Stefansson, Ambassador of the North*. Montreal: Harvest House, 1963.

This short biography is by a man whom Stefansson's Iago, Dr. Rudolph Anderson, called "Stefansson's press agent, advance agent, bill-peddler and yes-man." Considering all that, it's not a bad account! Yes, it is very pro-Stefansson and sketchy in some respects (it was boiled down from a much longer book that "Stef" failed to approve), but it contains much useful material, and it's interesting to get this glimpse of the "heroic" Stefansson alongside the figure constructed by the inevitable "qualifiers" who come later.

William Laird McKinlay. *Karluk: The Great Untold Story of Arctic Exploration*. London: Weidenfeld and Nicolson, 1976.

A terse, vivid account by a Scottish scientist and one of the survivors of the Karluk *disaster. McKinlay was writing a longer book on his traumatic* Karluk *experience when he died in 1983 at the age of ninety-four. This book is restrained, but seethes with anger at what McKinlay felt was Stefansson's careless handling of the expedition, his desertion of the ship, and other actions that the writer believed hardly earned the explorer his inflated reputation. The foreword by Magnus Magnusson, from a more objective perspective, supports McKinlay's estimate, without being altogether unfair to Stefansson.*

Richard J. Diubaldo. *Stefansson and the Canadian Arctic.* Montreal: McGill-Queen's University Press, 1978.

An academic study of Stefansson's relationship with Canada, and in particular an account of the fate of his ambitions and projects in the North. Its thesis is that Stefansson was an ambitious self-promoter, and an often erratic and ineffective planner and leader. This is not altogether untrue! The book is excellently detailed and researched but it tries too hard to be fair to Stefansson's civil service enemies (none of whom was a very attractive or interesting human being), and one of whom, Rudolph Anderson, was downright deranged on the subject of "Stef."

William R. Hunt. *Stef: A Biography of Vilhjalmur Stefansson, Canadian Arctic Explorer.* Vancouver: University of British Columbia Press, 1986.

This is the best general biography of Stefansson, well written, sympathetic, and yet careful and objective. Covers all aspects of Stefansson's life, without being exhaustive (or exhausting) in the irritable manner of some modern biographies. Indispensable.

Jennifer Niven. *The Ice Master: The Doomed 1913 Voyage of the Karluk*. New York: Hyperion, 2000.
An exciting book that tells the story of the Karluk *disaster. Stefansson is the villain (mostly offstage) and crusty old Captain Bartlett is "the ice master." It's a great story, and very well told — well-researched too — but since it deals with only one aspect of Stefansson's long career, some of the conclusions about him seem off-the-wall and very one-sided.*

Gísli Pálsson. *Travelling Passions: The Hidden Life of Vilhjalmur Stefansson*. (Translated from the Icelandic by Keneva Kunz). Winnipeg: University of Manitoba Press, 2003.
A marvellous addition to the basic information on Stefansson, including much new information on his connections with women, his Inuit family, his politics, and his time in Iceland and New York, and offering a modern anthropologist's take on Stefansson's achievements. Very readable, and extremely well balanced. Indispensable. See also Pálsson's brief but helpful web essays on Stefansson at www.thearctic.is/articles/topics/legacystefansson/. In the few places I draw on these in my text they are indicated by a simple "Arctic Web" ascription.

Other Works Consulted

Margaret Atwood. *Survival*. Toronto: Anansi, 1972.

Beowulf. Translated by Burton Raffel. New York: New American Library, 1963.

Edwin Bernbaum. *Shambhala: A Search for the Mythical Kingdom Beyond the Himalayas.* Boston and London: Shambhala, 2001.

George Boas. *Essays on Primitivism and Related Ideas in the Middle Ages.* New York: Octagon Books, 1966.

Brian Branston. *The Lost Gods of England.* London: Thames and Hudson, 1957.

Richard E. Byrd. *Alone.* New York: G.P. Putnam, 1938.

Richard E. Byrd. *Little America.* New York: G.P. Putnam, 1930.

H.R. Ellis Davidson. *Gods and Myths of Northern Europe.* Harmondsworth, UK: Penguin, 1964.

Jacqueline Decter. *Nicholas Roerich: The Life and Art of a Russian Master.* Rochester, VT: Park Street Press, 1989.

Dreams of a Summer Night. Scandinavian Painting at the Turn of the Century. London: Arts Council of Great Britain, 1986.

René Dubos. *A God Within.* New York: Scribner, 1973.

H. Essame. *Patton as Military Commander.* London: B.T. Batsford, 1973.

Robin Fedden. *English Travellers in the Near East.* London: The British Council, 1958.

William W. Fitzhugh and Elizabeth I. Ward. *Vikings: The North Atlantic Saga*. Washington, D.C.: Smithsonian Institution, 2000.

Joan Halifax. *Shaman: The Wounded Healer*. London: Thames and Hudson.

Tom Henighan. *Natural Space in Literature*. Ottawa: Golden Dog Press, 1982.

Tom Henighan "A Few Thoughts on the Northern Imagination," a talk delivered at the Canadian Museum of Civilization, in 2002, as part of the presentations accompanying the museum's Viking Exhibition, an expanded version of the Smithsonian exhibit entitled "Vikings: The North Atlantic Saga."

Thor Heyerdahl and Christopher Ralling. *Kon-Tiki Man: An Illustrated Biography of Thor Heyerdahl*. Vancouver and Toronto: Douglas and McIntyre, 1991.

Johannes V, Jensen. *The Long Journey*. New York: Alfred A. Knopf, 1945.

William Kurelek. *The Last of the Arctic*. Toronto: Pagurian Press, 1976.

Alfred Lansing. *Endurance: Shackleton's Incredible Voyage*. New York: McGraw-Hill, 1959.

Jack London. *The Call of the Wild*. 1903.

Alexander Maitland. *Wilfred Thesiger: The Life of the Great Explorer.* New York: Harper Perennial, 2007.

Desmond Morton. *A Short History of Canada.* Edmonton: Hurtig, 1983.

David Mountfield. *A History of Polar Exploration.* New York: The Dial Press, 1974.

Roald Nasgaard. *The Mystic North.* Toronto: University of Toronto and Art Galley of Ontario, 1984.

Mark Nuttal, ed. *Encyclopedia of the Arctic.* London: Routledge, 2004.

Knud Rasmussen. *Intellectual Culture of the Hudson Bay Eskimos.* Copenhagen: Gyldendal, 1930.

Edward Rice. *Captain Sir Richard Francis Burton.* New York: Harper Perennial, 2001.

Nicholas Roerich. *Heart of Asia.* New York: Roerich Museum Press, 1930.

Norman Smith, ed. *The Unbelievable Land.* Ottawa: The Queen's Printer, 1964.

J.R.R. Tolkien. "*Beowulf,* the Monsters and the Critics." Proceedings of the British Academy XXII (1936), 245–95.

Richard Trench. *Arabian Travellers: The European Discovery of Arabia*. London: Macmillan, 1986.

Victor F. Valentine and Frank G. Valee, eds. *Eskimo of the Canadian Arctic*. Toronto: McClelland & Stewart, 1968.

Laurens van der Post. *The Lost World of the Kalahari*. New York: William Morrow, 1958.

Index

Numbers in *italics* indicate pages of photographs.